Teacher's Guide & Test Bank to Accompany

ASTRONOMY:

From the Earth to the Universe

third edition

Jay M. Pasachoff

Williams College
Hopkins Observatory
Williamstown, MA

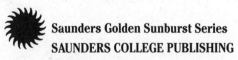 **Saunders Golden Sunburst Series**
SAUNDERS COLLEGE PUBLISHING
Philadelphia   New York   Chicago
San Francisco   Montreal   Toronto
London   Sydney   Tokyo

Address orders to: Saunders College Publishing--Marketing
383 Madison Avenue
New York, NY 10017

With a copy to: Jay M. Pasachoff
Williams College
Hopkins Observatory
Williamstown, MA 01267

Address Editorial Correspondence to:
Saunders College Publishing
Attn: John Vondeling, Editor
210 West Washington Square
Philadelphia, PA 19105

Editor: Elizabeth Stell

Teacher's Guide and Test Bank for
ASTRONOMY: FROM THE EARTH TO THE UNIVERSE, 3rd Edition.
ISBN
For use with
ASTRONOMY: FROM THE EARTH TO THE UNIVERSE, 3rd Edition.
ISBN 0-03-008114-9

Last digit is the print number: 9 8 7 6 5 4 3 2 1

CBS COLLEGE PUBLISHING
Saunders College Publishing
Holt, Rinehart and Winston
The Dryden Press

I am the very model of a modern-day astronomer.
I know all the absorption lines of inorganic monomer.
I won't confuse a spiral galaxy with an elliptical.
It puzzles me why anyone thinks CCD's are cryptical.
I understand the several tests of general relativity.
I've memorized the value of the solar emissivity.
I'll explicate the methods that are used for looking back eons
And tell you why the missing mass cannot be found in tachyons.

I've written several papers on the clustering of galaxies.
I see far more in black holes than an oversexed smart-alec sees.
I can discuss inflation with no thoughts of the economy,
For I am thoroughly versed in all of modern-day astronomy.

I know techniques for microwave and radio and gamma-ray.
I program my computer for its very largest RAM array.
I'm good at electronics, both the analog and digital.
I know a giant star will die much sooner than a midget'll.
I'm very well familiar with effects of the ionosphere.
I know why Oort disturbances might stimulate iguanas' fear.
I know redshifted starlight from the photons heated rocks emit.
If I worked on it, Hubble's constant wouldn't be approximate!

I've scanned the sun's companions out from Mercury to Nereid.
I've helped to find a pulsar with a millisecond period.
If scientists had fan clubs, then the masses would hosanna me
For all my contributions into modern-day astronomy.

Should you and I some dark night view the stellar cornucopia,
Don't ask me "Where's the Milky Way?" or "Is that Cassiopeia?"
Don't ask me where the Dippers are or if that's Mars or Sirius,
Because, if you don't know, to me it's equally mysterious!
I grew up in New York, so of the sky I know a smattering—
I've seen the sun and moon, and on the best days, Rayleigh scattering.
If I lived ninety years ago, I couldn't do as well as cope
For all I see are jiggly blobs when I look through a telescope.

To find the stars and planets, and the choice celestial scenery
I'm totally dependent on the latest in machinery.
Still, mispronouncing people call and ask where Comet Halley is,
So, seven months ago, I started living under alias!

Lloyd Rawley

PREFACE

This Teacher's Guide and Test Bank provides a complete set of aids for use with the text ASTRONOMY: FROM THE EARTH TO THE UNIVERSE, 3rd edition, by Jay M. Pasachoff.

The Teacher's Guide begins with sample course outlines for one-quarter, one-semester, or two-semester courses. The section on films includes lists of many astronomical films, where to buy, rent, or borrow them, and transcripts of and notes on several outstanding films. Students always respond well to films, but supplying them with a transcript or notes increases their comprehension. Suppliers of other audiovisual aids are also listed. I now recommend videodiscs for their flexibility and ease of use. At Williams College, we installed a projection television in our large physics/astronomy lecture room this year, and I have been enjoying the ease of projecting stills and movies from the Astronomy videodisc I list inside.

The section on laboratory exercises not only includes lists of commercially available exercises but also presents a number of full labs. Unlike most labs that are available in commercial laboratory manuals, which often dwell on details of observing, these labs are of a type that aid in the comprehension of text material. (More information on observing is provided in my book A Field Guide to the Stars and Planets, Peterson Field Guide Series, Houghton Mifflin, 1983, a joint publication with the late Donald H. Menzel.)

Next, we supply answers to all questions from the text. The rest of the book is devoted to a Test Bank containing sets of examinations. The exams include ten-minute quizzes (optional), one-hour exams, and final exams for one-semester courses. An exam featuring multiplechoice answers and fill-ins rather than essays and short answers is also supplied. Answers for all exams are given.

I wish to thank Elizabeth Stell for her excellent work on this book.

Please write me if you have any comments or corrections.

Jay M. Pasachoff

Williams College--Hopkins Observatory
Williamstown, MA   01267

TABLE OF CONTENTS

COURSE OUTLINES

This section provides a number of alternate outlines for one-quarter, one-, and two-semester courses. Also included are suggestions for films to accompany lectures. A listing of the various course outlines follows:

COURSE OUTLINES

Plan I: One-Quarter Course
(24 lectures)

(omit sections starred in text)

Note 1: Chapter 21, "The Sun," could also be taught between Chapters 6 and 7, or Chapters 17 and 18.

Note 2: Chapter 4 on "Light and Telescopes" can be taught between Chapters 1 and 2 or Chapters 17 and 18.

Note 3: Chapter 5 on observing, "The Sky and the Calendar," can be taught earlier or later.

Note 4: You could also choose to limit the course to either the solar system or stars and galaxies.

Lecture Number	
1	Introduction Chapter 1
2	History of Astronomy Chapters 2 and 3
3	Light and Telescopes Chapter 4
4	Observing, Magnitudes, Coordinates Chapter 5
5	The Structure and Origin of Our Solar System Chapter 6
6	The Earth and the Moon Chapter 7
7	Mercury and Venus Chapters 8 and 9
8	Mars Chapter 10
9	Jupiter Chapter 11
10	Saturn and the Outer Planets Chapters 12, 13 and 14
11	Comets, Meteorites, and Asteroids Chapters 15 and 16 and Life in the Universe Chapter 17
12	Hour Exam

COURSE OUTLINES

13   Ordinary Stars
      Chapter 18

14   Stellar Distances and Motions
      Chapter 19

15   Doubles, Variables, and Clusters
      Chapter 20

16   The Sun: A Star Close Up
      Chapter 21

17   Young and Middle-Aged Stars
      Chapter 22

18   The Death of Stars Like the Sun
      Chapter 23

19   Supernovae, Pulsars, and Other Neutron Stars
      Chapters 24 and 25

20   Stellar Black Holes
      Chapter 26

21   Structure of Our Galaxy
      Chapter 27
          and
      The Interstellar Medium
      Chapter 28

22   Galaxies
      Chapter 29

23   Quasars
      Chapter 30

24   Cosmology
      Chapter 31
          and
      The Past and Future of the Universe
      Chapter 32

COURSE OUTLINES

Plan IA: One-Quarter Course: Stars Expanded
(24 lectures)

(omit sections starred in text)

Note 1: Chapter 21, "The Sun," could also be taught between Chapters 6 and 7, or Chapters 17 and 18.

Note 2: Chapter 4 on "Light and Telescopes" can be taught between Chapters 1 and 2 or Chapters 17 and 18.

Note 3: Chapter 5 on observing, "The Sky and the Calendar," can be taught earlier or later.

Note 4: You could also choose to limit the course to either the solar system or stars and galaxies.

Lecture
Number

1   Introduction and the History of Astronomy
Chapter 1 and Section 2.1 - 2.5

2   Copernicus to Newton
Sections 2.5 - 2.6 and Chapter 3
   and
The Spectrum
Sections 4.1 and 4.2

3   Introduction to the Planets
Chapter 6

4   The Earth and the Moon
Chapter 7

5   Mercury, Venus, and Mars
Chapters 8, 9, and 10

6   Jupiter, Saturn, Uranus, Neptune, and Pluto
Chapters 11, 12, 13, and 14

7   Comets, Meteorites, and Asteroids
Chapters 15 and 16
   and
Life in the Universe
Chapter 17

8   Telescopes
Chapter 4

9   Observing: The Sky and the Calendar
Chapter 5

10   Ordinary Stars
Chapter 18

11   Stellar Distances and Motions
Chapter 19

COURSE OUTLINES

12 Doubles, Variables, and Clusters
    Chapter 20

13 Hour Exam

14 The Sun: A Star Close Up
    Chapter 21

15 Young and Middle-Aged Stars
    Chapter 22

16 The Death of Stars Like the Sun
    Chapter 23

17 Supernovae, Pulsars, and Other Neutron Stars
    Chapters 24 and 25

18 Stellar Black Holes
    Chapter 26

19 Structure of Our Galaxy
    Chapter 27

20 The Interstellar Medium
    Chapter 28

21 Galaxies
    Chapter 29

22 Quasars
    Chapter 30

23 Cosmology
    Chapter 31

24 The Past and Future of the Universe, Review, and Summary
    Chapter 32

COURSE OUTLINES

Plan II: One-Trimester Course
(30 lectures)

Note 1: Chapter 21, "The Sun," could also be taught between Chapters 6 and 7, or Chapters 17 and 18.

Note 2: Chapter 4 on "Light and Telescopes" can be taught between Chapters 1 and 2 or Chapters 17 and 18.

Note 3: Chapter 5 on observing, "The Sky and the Calendar," can be taught earlier or later.

Lecture
Number

1   Introduction
      Chapter 1

2   The Development of Astronomy
      Chapters 2 and 3

3   Light and Telescopes
      Chapter 4

4   The Sky and the Calendar
      Chapter 5

5   The Structure and Origin of Our Solar System
      Chapter 6

6   Our Earth
      Section 7.1

7   The Moon
      Chapter 7

8   Mercury
      Chapter 8

9   Venus
      Chapter 9

10   Mars
      Chapter 10

11   Jupiter
      Chapter 11

12   Saturn
      Chapter 12

13   The Outer Planets
      Chapters 13 and 14

14   Comets, Meteorites, and Asteroids
      Chapters 15 and 16

COURSE OUTLINES

15   Hour Exam

16   Life in the Universe
      Chapter 17

17   Ordinary Stars
      Chapter 18

18   Stellar Distances and Motions
      Chapter 19

19   Doubles, Variables, and Clusters
      Chapter 20

20   The Sun: A Star Close Up
      Chapter 21

21   Young and Middle-Aged Stars
      Chapter 22

22   The Death of Stars Like the Sun
      Chapter 23

23   Supernovae, Pulsars, and Other Neutron Stars
      Chapters 24 and 25

24   Stellar Black Holes
      Chapter 26

25   Structure of Our Galaxy
      Chapter 27

26   The Interstellar Medium
      Chapter 28

27   Galaxies
      Chapter 29

28   Quasars
      Chapter 30

29   Cosmology
      Chapter 31

30   The Past and Future of the Universe
      Chapter 32

COURSE OUTLINES

Plan III: One-Semester Course
(36 lectures)

Note 1: Chapter 21, "The Sun," could also be taught between Chapters 6 and 7, or Chapters 17 and 18.

Note 2: Chapter 4 on "Light and Telescopes" can be taught between Chapters 1 and 2 or Chapters 17 and 18.

Note 3: Chapter 5 on observing, "The Sky and the Calendar," can be taught earlier or later.

Note 4: You could also choose to limit the course to either the solar system or stars and galaxies.

Lecture
Number

1   Introduction
     Chapter 1

2   The History of Astronomy
     Chapter 2

3   Kepler, Galileo, Newton, & Halley
     Chapter 3

4   Light and Telescopes
     Sections 4.1 - 4.6

5   Telescopes
     Sections 4.7 - 4.14

6   The Sky and the Calendar
     Chapter 5, Appendix 10

7   The Structure and Origin of the Solar System
     Chapter 6

8   Our Earth and Moon
     Chapter 7

9   Mercury
     Chapter 8

10   Venus
     Chapter 9

11   Mars
     Chapter 10

12   Hour Exam

COURSE OUTLINES

13   Jupiter
      Chapter 11

14   Saturn
      Chapter 12

15   The Outer Planets
      Chapters 13 and 14

16   Comets, Meteorites, and Asteroids
      Chapters 15 and 16

17   Life in the Universe
      Chapter 17

18   Stars: Colors and Spectral Lines
      Sections 18.1 - 18.3

19   Stellar Spectra
      Sections 18.4 - 18.5

20   H-R Diagrams
      Sections 19.1 - 19.4

21   Stellar Motions
      Sections 19.5 - 19.6

22   Doubles, Variables, and Clusters
      Chapter 20

23   The Quiet Sun
      Sections 21.1 - 21.5

24   The Active Sun, the Solar Constant, and Relativity
      Sections 21.6 - 21.10

25   Hour Exam

26   Young and Middle-Aged Stars
      Chapter 22

27   The Death of Stars Like the Sun
      Chapter 23

28   Supernovae, Pulsars, and Other Neutron Stars
      Chapters 24 and 25

29   Stellar Black Holes
      Chapter 26

30   Structure of Our Galaxy
      Chapter 27

31   The Interstellar Medium
      Chapter 28

COURSE OUTLINES

Plan IV: One-Semester Course
The Solar System
(24 lectures)

Note: 24 lectures are based on a format of 2 lectures a week for 12 weeks. If your semester includes more than 24 lectures, you can expand this course outline most easily with the use of the films listed later.

Lecture
Number

1   Introduction
        Chapter 1

2   The Development of Astronomy
        Chapter 2

3   Kepler, Galileo, Newton, and Halley
        Chapter 3

4   Light and Telescopes
        Chapter 4

5   The Sky and the Calendar
        Chapter 5, Appendix 10

6   The Structure and Origin of the Solar System
        Chapter 6

7   Our Earth
        Section 7.1

8   The Moon
        Sections 7.2a - 7.2d, film

9   The Moon
        Sections 7.3b - 7.3

10   The Quiet Sun
        Sections 21.1 - 21.5

11   The Active Sun, the Solar Constant, and Relativity
        Sections 21.6 - 21.10

12   Hour Exam

13   Mercury
        Chapter 8

14   Venus
        Chapter 9

COURSE OUTLINES

15 Mars
    Chapter 10, film

16 Jupiter
    Sections 11.1 - 11.4, film

17 Jupiter's Moons and Saturn's Rings
    Sections 11.5 - 12.2a

18 Saturn
    Chapter 12

19 Uranus, Neptune, and Pluto
    Chapters 13 and 14

20 Halley's and Other Comets
    Chapter 15

21 Meteorites and Asteroids
    Chapter 16

22 Life in the Universe
    Sections 17.1 - 17.3

23 Life in the Universe
    Sections 17.4 - 17.6

24 The Past and Future of the Universe
    Chapter 32

COURSE OUTLINES

Plan V: One-Semester Course
Stars, Galaxies, and Beyond
(24 lectures)

Note: 24 lectures are based on a format of 2 lectures a week for 12 weeks. If your semester includes more than 24 lectures, you can expand this course outline most easily with the use of the films listed later.

Lecture
Number

1   Introduction
     Chapter 1

2   Light and Observing
     Chapters 4 and 5 (or appropriate review)

3   Ordinary Stars
     Sections 18.1 - 18.3

4   Stellar Spectra
     Sections 18.4 - 18.5

5   H-R Diagrams
     Sections 19.1 - 19.4

6   Stellar Motions
     Sections 19.5 - 19.6

7   Doubles, Variables, and Clusters
     Chapter 20

8   The Sun: A Star Close Up
     Chapter 21

9   Stellar Evolution
     Sections 22.1 - 22.3

10   Main-Sequence Stars and the Neutrino Experiment
     Sections 22.4 - 22.7

11   Red Giants, Planetary Nebulae, White Dwarfs, and Novae
     Chapter 23

12   Supernovae
     Chapter 24

13   Pulsars and Other Neutron Stars
     Chapter 25

COURSE OUTLINES

COURSE OUTLINES

Plan VI: One-Semester Course
Planets and Stars

Note: 24 lectures are based on a format of 2 lectures a week for 12 weeks. If your semester includes more than 24 lectures, you can expand this course outline most easily with the use of the films listed later.

Lecture
Number

1   Introduction
       Chapter 1

2   The Development of Astronomy
       Chapter 2

3   Kepler, Galileo, Neton, and Halley
       Chapter 3

4   Light and Telescopes
       Chapter 4

5   The Sky
       Sections 5.1 - 5.4

6   The Structure and Origin of the Solar System
       Chapter 6

7   Our Earth
       Section 7.1

8   The Moon
       Section 7.2

9   Mercury
       Chapter 8

10   Venus
       Chapter 9

11   Hour Exam

12   Mars
       Chapter 10

13   Jupiter
       Chapter 11

14   Saturn and the Outer Planets
       Chapters 12, 13 and 14

COURSE OUTLINES

15   Comets, Meteorites, and Asteroids
Chapters 15 and 16

16   Life in the Universe
Chapter 17

17   Ordinary Stars
Chapter 18

18   Stellar Distances and Motions
Chapter 19

19   Doubles, Variables, and Clusters
Chapter 20

20   The Sun: A Star Close Up
Chapter 21

21   Young and Middle-Aged Stars
Chapter 22

22   The Death of Stars Like the Sun
Chapter 23

23   Supernovae, Pulsars, and Other Neutron Stars
Chapters 24 and 25

24   Stellar Black Holes
Chapter 26

COURSE OUTLINES

Plan VII: One-Semester Course
Our Galaxy and Beyond

Note: 24 lectures are based on a format of 2 lectures a week for 12 weeks. If your semester includes more than 24 lectures, you can expand this course outline most easily with the use of the films listed later.

Lecture
Number

1   Introduction
     Chapter 1

2   Distance Scale, The Sky
     Section 31.4; Appendix 10; film: Exploring the Milky Way

3   Nebulae
     Section 27.1

4   Structure of Our Galaxy
     Sections 27.2 - 27.3

5   High-Energy Astronomy and Spiral Structure
     Sections 27.4 - 27.5

6   Observing
     Chapter 5; Appendix 10; film: The Universe from Palomar

7   The Interstellar Medium
     Sections 28.1 - 28.2

8   Radio Astronomy
     Sections 28.3 - 28.4; film: The Invisible Universe

9   Spectral-Line Radio Astronomy and Star Formation
     Sections 28.5 - 28.9

10   Classifying Galaxies
     Section 29.1; film: Realm of the Galaxies

11   Galaxies: Clusters and the Missing-Mass Problem
     Sections 26.2 and 26.3

12   Hour Exam

13   Galaxies: Expansion of the Universe
     Section 29.4

14   Radio Galaxies
     Section 29.5

15   Interferometry
     Section 29.6

16   Hour Exam

17   Quasars
      Sections 30.1 - 30.4

18   Quasars
      Sections 30.5 - 30.9

19   Cosmology: Olbers's Paradox and the Big Bang
      Sections 31.1 - 31.2

20   Cosmology: Steady State
      Section 31.3

21   Background Radiation
      Section 31.4; film: Three Degrees

22   Origin of the Elements
      Section 32.1; film: The Origin of the Elements

23   The Future of the Universe
      Sections 32.2 - 32.4

24   Review and Summary

COURSE OUTLINES

Plan VIII: Two-Semester Course
First Semester: Planets and Stars
Second Semester: Our Galaxy and Beyond

For a two-semester course, combine the previous two course outlines, using Plan VI: Planets and Stars (24 lectures) for the first semester and Plan VII: Our Galaxy and Beyond (24 lectures) for the second semester.

If your semester includes more than 24 lectures, you can expand these course outlines most easily with the use of the films listed later.

COURSE OUTLINES

A Sample Selection of Films

Miscellaneous

PF	Powers of Ten
NOAO	Journey into Light
NOAO	Skies of the Andes

Planets

	Copernicus
PF	Planetary Motion and Kepler's Laws
H-M	Apollo 17: On the Shoulders of Giants
NASA	The Moon: A Giant Step in Geology
EB	Mercury/Exploration of a Planet
NASA	Mars: Chemistry Searches for Life
MLA	Planet Mars
NASA	Resolution on Saturn
C/MTI	The Planet That Got Knocked on its Side

Stars and Galaxies

NOAO	Stars, Galaxies, and Southern Skies
MLA	Exploring the Milky Way
H-M	The Doppler Effect in Sound and Light
H-M	Star System "Xi Ursae Majoris"
H-M	Algol, the Demon Star
NASA	Skylab and the Sun
T-L	Birth and Death of a Star
H-M	Sirius and the White Dwarf
FI	Crab Nebula
CAL	The Universe from Palomar
MTP	Three Degrees
HP	The Universe: Man's Changing Perceptions
NRAO	The Invisible Universe

Ordering information for these and many other films is listed in the following section.

AUDIOVISUAL AIDS

FILMS

Note: Abbreviations indicate where to rent or buy films. Distributors for the films are listed later in this section. All films are in color (16mm) unless otherwise noted. We have verified availability and ordering information in the spring of 1986.

General

PF   Powers of Ten
9 minutes. Revised 1978. Rental Fee: $55.
Purchase Price: 16mm $250, video $250.
(see also videodisc section)
> A cosmic zoom view of our universe in powers of 10, up to clusters of galaxies and then down to an atomic nucleus. The revised film is in color and is narrated by Philip Morrison. Its coverage of the microscopic world is improved from that of the first edition. Some viewers may prefer the original 1968 version for a variety of reasons. One contrast is the informal tone of Morrison's narration versus the flat, clear enunciation of the narrator of the original version. A film by the Office of Charles and Ray Eames.

CAL
IU   The Universe from Palomar
30 minutes. 1967. Rental Fee: CAL $25 for one-time showing, $10 each additional showing; IU $12.15.
Purchase Price: CAL $250; IU $250.
> A discussion of the 5-meter telescope, the scientists who work with it, and some results of their work.

NASA   Universe
28 minutes. 1976. HQ220.
> An overview of the range of astronomical objects. Although the film has won prizes, it contains several scientific errors.

NOAO   Journey into Light
28 minutes. 1978. No rental fee, but $12 postage and handling fee must be submitted with order.
Purchase Price: $250.
> The current film about the Kitt Peak National Observatory and research there.

NOAO   Skies of the Andes
28 minutes. 1972. No rental fee, but $12 postage and handling fee must be submitted with order.
Purchase Price: $250.
> About the site selection and construction of the Cerro Tololo Inter-American Observatory in Chile. Too much introduction, but about 15 minutes of worthwhile material.

FILMS

NOAO   The Observatories
MTPS   27 minutes. 1982. NOAO: No rental fee, but $12 postage
        and handling fee must be submitted with order; MTP:
        Rental is cost of return postage only.
        Purchase Price: NOAO $250.
          A tour of the six major observatories supported by the
          NSF: CTIO, NSO, Green Bank, VLA, Arecibo, Kitt Peak.
          Produced by the NSF.

MG   Cosmology Before Newton
        22 minutes. 1982. Rental Fee: $40.
        Purchase Price: 16mm $495, video $249.
          A gentle film about historical astronomy from the Open
          University in Britain. Very well done.

MAN   A Brief History of Astronomy
        10 minutes. 1970. Purchase Price: $150.
          Traces the development of astronomy from primitive man
          to modern space exploration. Covers Egyptians, Greeks,
          Copernicus, Galileo, etc. Produced by the Manitoba
          Museum of Nature and Man.

UA   Mirrors on the Universe: The MMT Story
        28 minutes. 1979. Rental Fee: $13.50.
        Purchase Price: $385; 3/4" U-matic $275.
          A joint production of the University of Arizona and the
          Smithsonian Institution that documents the construction
          of the MMT.

MTPS   Images of Einstein
        12 minutes. Rental Fee: cost of return postage only.
          Explores the interactions between artistic and
          scientific creativity, including discussion of Albert
          Einstein.

NASA   Space Research and Your Health
        15 minutes. 1981. HQ319.
          Examines how NASA research has contributed to a variety
          of medical advances. A bit elementary.

NASA   Space Research and Your Transportation
        15 minutes. 1981. HQ320.
          Examines how NASA research has contributed to techno-
          logical improvements in land, sea, air, and space
          travel. A bit elementary.

NASA   Space Research and Your Home and Environment
        15 minutes. 1981. HQ321.
          Examines how NASA research contributes to a better
          understanding of the earth, protection of its resources,
          and energy-saving technologies. A bit elementary.

FI   Cosmos
        For rental, purchasing, or licensing of Carl Sagan's
        Cosmos series, write or telephone Films, Inc.

FILMS

FI   **Journey to the Stars**
16 minutes. Rental Fee: $45.
Purchase Price: 16mm $315, video $205 (3/4" or 1/2").
A mixture of old and new footage, starting with the Hale telescope, the sun, and a nice view from space of the earth's phases. The opening, "This will be the longest journey that any of us ever take--a trip through the mountains to an astronomy observatory and then a voyage through the stars to the boundaries of space and time," sets the tone and level, which is perhaps too elementary for college students. A theme of the movie is stellar evolution and whether it provides enough time for different spectral types or for binaries to allow for life to evolve. The animation of the orbits of the members of a binary is wrong. Nice features include the Gerola and Seiden movie of spiral structure forming as simulated with a computer at IBM, and footage of Penzias and Wilson with their telescope moving.

BF
KSU   **Astronomy: The Cosmic Quest**
26 minutes. 1981. Rental Fee: BF $58.50 plus shipping; KSU $15.74.
Purchase Price: BF $585 plus shipping.
Explores the frontiers of astronomy through the work and thoughts of 4 leading astronomers.

PBS   **Spaceflight**
4 1-hour programs. 1985. Rental Fee: $95 per program.
Purchase Price: $250 per program; $800 for all 4.
The history of spaceflight. The first program takes us up to Sputnik. The second and third programs focus on the manned American missions, up to the Apollo-Soyuz linking in space. The last discusses "Star Wars," the Space Telescope, and other future projects.

PBS   **The Creation of the Universe.**
90 minutes. 1985. Rental Fee: $125.
Purchase Price: $400.
A wonderful television film that won the 1985 AAAS-Westinghouse Science Journalism Award. It was written and narrated by Timothy Ferris. The first hour discusses fundamental particles and forces; the second part looks back in time to the early universe. A transcript appears later in this Teacher's Guide. Teacher's Manual and handouts are available fromm TelEd, Inc., 7449 Melrose Ave., Los Angeles, CA 90046.

BFF   **The Sun Dagger**
30 and (60) minute versions. Rental Fee: $50 ($85).
Purchase Price: 16 mm $550 ($850); video $250 ($450).
Archaeoastronomy: A film about the sun dagger in Chaco Canyon.

C/MTI   **What Einstein Never Knew**
58 minutes. 1986. Rental Fee: $125.

FILMS

Purchase Price: $250 (video only; all formats).
About particle physics and fundamental forces.
Transcript available for $4.00 from WGBH Transcripts,
P.O. Box 322, Boston, MA 02134. Teacher's Guide
available for $2.00 from WGBH NOVA Teacher's Guide, Box
222, South Easton, MA 02735.

FILMS

Extraterrestrial Life

NASA   Life Beyond Earth and the Mind of Man
27 minutes. 1975.
Purchase Price: $265.00.
An edited version of a symposium held at Boston University in 1972 to explore the implications of the existence of extraterrestrial life.

EAV   Life in the Universe
1977. Available for purchase only: $231 for 6 cassettes and 6 filmstrips (or 6 LP's), teacher's notes, full text of narration, suggestions for use, topics for discussion, and bibliography. Automatic and manual projection.
Presents current scientific speculation basic to space exploration, and particularly to the search for life. Students are introduced to current theories regarding the origin of the universe, the origin of galaxies, the origin of the solar system, the origin of life itself. Based on a permanent exhibition at the Smithsonian Institution's National Air and Space Museum.

NASA   Who's Out There?
28 minutes. 1975. HQ226.
Narrated by Orson Welles.

NASA   Life?
15 minutes. 1976. HQ261.
A nice discussion of life and its definition, with no astronomical connection except for a pre-launch mention of Viking at the end.

C/MTI   Is Anybody Out There?
58 minutes. 1986. Rental Fee: $125.
Purchase Price: $250. (Video only; no 16-mm version)
A NOVA report on SETI, narrated by Lily Tomlin, who does well, and Ernestine, who carries on too long. Features Frank Drake, Carl Sagan, Paul Horowitz, Ben Zuckerman, Rich Terrile, and others. Nice sections on the Drake equation and on the difficulties of finding appropriate media of communication. Transcript available for $4.00 from WGBH Transcripts, P.O. Box 322, Boston, MA 02134 (specify title and date shown: 11/18/86).

FILMS

Skylab and Apollo-Soyuz Films

NASA   Opportunities in Zero-Gravity
28 minutes. 1976.
> An overview of the Skylab Science Demonstration Series. Scientist/astronaut Owen Garriott narrates. See also:

Zero-G
15 minutes. 1974. HQa260A.
> Introduction to weightlessness, including scenes of people floating. Filmed on Skylab.

Gyroscopes in Space
15 minutes. 1974. HQa260C.
> Dr. Garriott discusses the Skylab experiment and illustrates the basic principles of gyroscopes by using kinescope and other film footage from Skylab.

Fluids in Weightlessness
15 minutes. 1974. HQa260D.
> Dr. Garriott discusses the behavior of fluids in orbit, using videotape from Skylab.

Magnetism in Space
20 minutes. 1975. HQa260E.
> Dr. Garriott discusses the behavior of magnets in orbit. Filmed on Skylab.

Magnetic Effects in Space
14 minutes. 1975. HQa260F.
> Dr. Garriott uses a classroom format to teach about magnetism, with Skylab videotape for illustrations.

Skylab: The First 40 Days
25 minutes. 1973.

Skylab: The Second Manned Mission--A Scientific Harvest
37 minutes. 1974.
> Includes discussions of Skylab's solar research.

Skylab: Space Station 1
28 minutes. 1974.

Spaceship Skylab: Wings of Discovery
28 minutes. 1974.

The Mission of Apollo-Soyuz
28 minutes. 1975. HQ 256.

FILMS

The Stars

T-L   Birth and Death of a Star
35 minutes. Rental Fee: $40.
Purchase Price: $400.
> An excellent film about the origin, evolution, and death processes of stars. Stellar structure and nuclear reactions are also discussed. A transcript appears later in this Teacher's Guide.

IU
SU
FI
KSU   Crab Nebula
56 minutes. 1972. Rental Fee: IU $24; KSU $30.50; SU $32; FI $75.
Purchase Price: FI 16mm $750; video 1/2" $198, 3/4" 298.
> An excellent film from the Public Broadcasting System's NOVA series, about many aspects of the Crab Nebula, including a good discussion on the discovery of pulsars. The last 10 minutes can be omitted. Notes about the film appear later in this Teacher's Guide.

H-M   The Star System "Xi Ursae Majoris"
Sirius and the White Dwarf
Algol, the Demon Star
The Motions of Stars
Star Clusters
Approximately 8 minutes each. Rental Fee: $35 each.
Purchase Price: $250 each.
> The images in these films are generated by a computer to demonstrate each concept; electronic music and narration are provided. A pamphlet, entitled "Film Notes" by Michael Zeilik, II, is included.
> Super-8mm color film loops are available from Kalmia. Purchase Price: $34.50 each (discount for a set). For ordering information, write to: Kalmia Company, Inc., 21 West Circle, Concord, MA 01742.

NASA   Ultraviolet Stellar Astronomy
11 minutes. 1973.
Filmed on Skylab.

MG-H
KSU   Black Holes and Quasars
29 minutes. 1982. Rental Fee: MG-H $50; KSU $18.75.
Purchase Price: MG-H 16mm $515, video $395.
> Isaac Asimov and Mario Machado narrate. Ray Bradbury, the late George Abell, and others are featured briefly. Introduction is vague and graphics are too jazzy at the beginning. Much good material is presented but treated too broadly.

FILMS

MG   The Message of Starlight
KSU   24 minutes. 1983. Rental Fee: MG $40 (video only); KSU $15.75 (VHS).
Purchase Price: video, all formats, $198.
Stellar evolution and how it is studied. A program made for the British Open University, filmed at the Royal Greenwich Observatory, Herstmonceux, and at Oxford.

FILMS

The Sun

CAL   The Solar Atmosphere
30 minutes. 1970. Rental Fee: $20; $5 each additional showing.
Purchase Price: $200.
> An excellent film that includes not only a varied selection of time-lapse movies of solar phenomena but also descriptions of solar observing techniques and each of the major parts of the solar atmosphere. Suitable for all but the most elementary students.

BBSO   Big Bear Solar Observatory 1971 Show Film
Black-and-white. Purchase Price: approximately $60 plus shipping.
> Time-lapse solar movies.

BBSO   Big Bear Solar Observatory 1978 Show Film
Black-and-white. Purchase Price: approximately $60 plus shipping.
> Time-lapse solar movies with a sound cassette containing narration by Harold Zirin.

BBSO   Big Bear Solar Observatory: The Flares of August
Black-and-white. Purchase Price: approximately $60 plus shipping.
> Time-lapse solar movies of the flares of August 1972.

BBSO   Prominence Film
Purchase Price: approximately $120.
> Two 400-ft. rolls of movies of solar prominences.

MF   Explosions on the Sun
NCAR   9 minutes. 1945-46, reshot from original negatives in 1978. Rental Fee: MF $15 for 3 days.
Purchase Price: NCAR $175.
Black-and-white, silent.
> Spectacular time-lapse movies of intense solar activity three solar maxima ago.

MF   Shadow Across the Sun
NCAR   27 minutes. 1970. Rental Fee: MF $15 for 3 days.
Purchase Price: NCAR $275.
> The story of the High Altitude Observatory's expedition to southern Mexico to observe the 1970 total solar eclipse.

ISU   Day of the Dark Sun
17 minutes. 1973. Rental Fee: $14.40 for 3 days.
Purchase Price: $200.
> A report on the NSF 1973 eclipse expedition to Kenya.

NASA   Skylab and the Sun
13 minutes. 1977. Purchase Price: $140.00.

FILMS

KAR   Solar Eclipse '73
NG   24 minutes. 1974. Rental Fee: KAR $32 plus $3 shipping. Purchase Price: NG 16mm $345 plus $12.25 shipping, video $241 plus $9.25 shipping; KAR 16mm $345, video $241.

Film comes with associated printed materials and directions for an in-class experiment for construction of a small spectroscope and information about spectroscopes and Fraunhofer lines. One of each of the following comes with the film when purchased: Sun & Light--a booklet; a device to illustrate spectra of the sun and laboratory devices; Teacher's Guide; a piece of diffraction grating. Additional sets of the associated materials are available for $1.50 from National Geographic.

KAR   The Sun: The Earth's Star
NG   20 minutes. 1980. Rental Fee: KAR $32 plus $3 shipping. Purchase Price: KAR 16mm $355, video $248; NG 16mm $355 plus $12.25 shipping, video $248 plus $9.25 shipping.

Discusses features of the sun and how they affect the earth.

KSU   What is an Eclipse?
11 minutes. 1985. Rental Fee: $8.50.

Uses animation and 3-D models to illustrate causes of solar and lunar eclipses. Actual time-lapse, telescopic motion pictures of both a lunar and solar eclipse are shown, as well as new astronomical telescopes and recent discoveries about eclipses and the sun. Revised edition.

CS   Spherical Oscillation Paterns
6 minutes. 1984. (Los Alamos Film X-331)

A computer-simulated movie for classroom demonstrations of the spherical harmonics. An explicit demonstration of the time dependence of several $\underline{Y}_{lm}$'s.

--   Solar Oscillations
Approximately 3 minutes. 1985. Purchase Price: approximately $15 plus shipping (16mm only).

Martin Arrambide, Richard Faller and Robin Stebbins have generated an animated movie to demonstrate the sun's modes of oscillation. The movie shows the geometric character of some simple modes and the progression to the more complex geometric patterns. It also attempts to suggest the complexity of solar ocillations resulting from the superposition of many modes. The movie was made at Sacramento Peak Observatory for inclusion in the "Solar Sea" segment of the PBS documentary, "Planet Earth."

To order, contact:

Mr. Carl Posey   (602) 327-5511, x202
Public Information Officer
NOAO
P.O. Box 26732
Tucson, AZ 85726

FILMS

The Milky Way Galaxy

MLA
KSU

A Radio View of the Universe
29 minutes. 1967. Rental Fee: MLA $40; KSU $18.75.
Purchase Price: $580.

Shows an astronomer (Morton Roberts) planning a project and observing at the Agassiz Station of the Harvard College Observatory and at the National Radio Astronomy Observatory in Green Bank. The project involves using the 21-cm radio line of atomic hydrogen to determine the total hydrogen content of external galaxies. Produced by the American Astronomical Society.

MLA

Exploring the Milky Way
29 minutes. 1967. Rental Fee: $40. Purchase Price: $580.

Describes the use of Mira variables to deduce the structure and evolutionary history of our Milky Way Galaxy. Also shows an astronomer's (George Preston's) typical night's work at Lick Observatory. Produced by the AAS.

NRAO

The Invisible Universe
13 minutes. 1974. Purchase Price: approx. $100.
Radio astronomy and its new frontiers are examined in this delightful, sometimes humorous film. Produced by the National Radio Astronomy Observatory.

To order film, write to: Wallace R. Oref, Public Education Office, National Radio Astronomy Observatory, P.O. Box 2, Green Bank, WV 24944. Make purchase order or check payable to: Byron Motion Pictures, Inc., 65 K St., N.E., Washington, DC 20002.

Film can also be purchased directly from Colorlab, 5708 Arundale Ave., Rockville, MD 20852, (301) 770-2128. Price: $72.44 plus shipping.

CAL
IU

The Universe from Palomar
30 minutes. 1967. Rental Fee: CAL $25 one-time showing; $5 each additional showing; IU $12.15.
Purchase Price: CAL $250; IU $250.

A discussion of the 5-meter telescope, the scientists who work with it, and some of their results.

FILMS

Galaxies and Beyond

Three Degrees
25 minutes. 1979. Produced by Bell Laboratories. Rental Fee: $25 plus $3 shipping (16mm film or video).
Purchase Price: 16mm $334.14 plus $3 shipping, U-matic, BETA and VHS $74 plus $3 shipping.
Contact: John Francis, M. G. S., 619 W. 54th St., New York, NY 10019.
The story of the discovery by Penzias and Wilson.

MG-H The Origin of the Elements
KSU 18 minutes. 1973. Rental Fee: 16mm MG-H: $50, KSU: $8.50.
Purchase Price: 16mm $335, video $260.
Explores the nuclear and stellar processes involved in creating the elements; portrays life cycles of stars.

-- An Introduction to the Periodic Table
3 minutes. 1982. (video only)
Purchase Price: $35 (or $15 if you supply blank tape).
Tom Lehrer's own rendition of his song, "The Elements," lip-synched by a white coated actor in a chemistry lecture hall with a chart of the elements.
Available from:
Dr. Frank S. Lestingi
State University College
1300 Elmwood Ave.
Buffalo, NY 14222

HP The Universe: Man's Changing Perceptions
29 minutes. 1976. Rental Fee: $45.
Purchase Price: $475.
An excellent and scientifically accurate film. Makes a wonderful summary of the course. An outline of the film appears later in this Teacher's Guide.

IU Violent Universe
148 minutes. 1969. Rental Fee: $34.65.
Purchase Price: $750.
From a wonderful television series; unfortunately available only in black-and-white. A book by Nigel Calder, based on the movie, is available (1969).

NOAO Stars, Galaxies, and Southern Skies
30 minutes. 1973. No rental fee, but $12 postage and handling fee must be submitted with order.
Purchase Price: $250.
Emphasis on southern hemisphere studies of stars and galaxies. With the late Bart Bok at Cerro Tololo.

Galaxy Studies Through Computer Simulation
Copies of the actual scientific movies made by Prof. Richard H. Miller of the University of Chicago and colleagues. Copies will be made individually at a

FILMS

regular film laboratory and are thus expensive: about $0.15/sq. ft. for black-and-white (typical length: 400 ft.).

For a list of films available, write to: R. H. Miller and B. F. Smith, University of Chicago, Astronomy and Astrophysics Center, 5640 South Ellis Ave., Chicago, IL 60647.

T-L   Beyond the Milky Way
57 minutes. Rental Price: video, all formats, $85.
Purchase Price: 16mm not available, video, all formats, $250.
NOVA on modern astronomy.

PF   A Search for Antiworlds
1977. 25 minutes. Rental Price $60 plus 7.50 shipping.
Purchase Price: 16mm $425; video $395; add $3 for shipping with purchase.
A film about a search for antimatter, with participation by Hannes Alfvén and Luis Alvarez.

FILMS

The Solar System

T-L
KSU   The Starry Messenger
52 minutes. 1974. Rental fee: T-L $100; KSU $30.50.
No. 6 in Jacob Bronowski's "The Ascent of Man" series.
About the development and acceptance of the Copernican
theory, including a masterful part about the trial of
Galileo.

T-L
KSU   The Majestic Clockwork
52 minutes. 1974. Rental Fee: T-L $100; KSU $30.50.
No. 7 in Bronowski's "The Ascent of Man" series. The
film covers the contributions of Newton and Einstein and
the revolution that ensued when Einstein's theory of
relativity upset Newton's elegant description of the
universe.

KSU   The Heavens are Telling
25 minutes. 1973. Rental Fee: $15.75.
Cosmological speculation before 1600: conceptions of
the universe held by Babylonian astronomers, Plato and
the Pythagoreans, Eudoxus and Aristotle, and Copernicus.

PF   Copernicus
10 minutes. 1973. Rental Fee: $45.
Purchase Price: 16mm $295, video $295.
A look at surroundings, artifacts, and documents of
Copernicus and how he came to advance the heliocentric
theory. A Charles and Ray Eames film.

H-M   The Motions of Attracting Bodies
Planetary Motion and Kepler's Laws
The Doppler Effect in Sound and Light
Approximately 8 minutes each. Rental Fee: $35 each.
Purchase Price: $250.
The images in these films are generated by a computer to
demonstrate each concept; electronic music and narration
are provided. A pamphlet, entitled "Film Notes" by
Michael Zeilik, II, is included.
Super-8mm color film loops are available from Kalmia.
Purchase Price: $34.50 each (discount for set); the
first two films above come in two parts each. Write for
catalog and ordering information: Kalmia Company, Inc.,
21 West Circle, Concord, MA 01742.

KSU   Kepler's 3 Laws
The Kepler Problem
30 minutes. 1985. Rental Fee: 18.75.
Both from The Mechanical Universe, a series based on a
Caltech physics course; nearly all demonstrations are
accompanied by mathematical descriptions. How Kepler
developed his laws of planetary motions, and how Newton
solved "the Kepler problem"; also reviews Kepler's life
and work.

SU   The Planets

52 minutes. 1976. Rental Fee: $36 for 2 days.
An interesting film from the Public Broadcasting System's NOVA series. Produced by the BBC.

NASA   Our Solar System
5 minutes. 1973. HQ234.
Teaches names, line-up, and characteristics of the planets in our solar system. Animated--"My Very Educated Mother Just Served Us Nine Pizzas."

KAR   The Solar System
EB   20 minutes. 1980. Rental Fee: KAR $32 plus $3 shipping;
NG   EB $34.50
Purchase Price: EB 16mm $345, video $275; NG 16mm $355 plus $12.95 shipping, video $248 plus $9.25 shipping; KAR 16mm $355, video $248.
Shows spectacular views from recent spacecraft voyages of planets and some of the other bodies in the solar system.

MLA   Earth: Rotation and Revolution
8 1/2 minutes. Purchase Price: 16mm $88, video 80.
Reveals how the earth tilts both toward and away from the sun, using realistic earth and sun models. Produced by the AAS.

MLA   Moon: Motions and Phases
7 minutes. Purchase Price: 16 mm $88, video $80.
Via split-screen technique, simultaneously shows revolution of the moon about the earth and progression of lunar phases from earth. Produced by the AAS.

NASA   The Moon--An Emerging Planet
13 minutes. 1973. HQa230.
Developing a lunar chronology.

NASA   Reading the Moon's Secrets
18 minutes. 1976. HQa275.
Too elementary a style of presentation, with a schoolteacher asking questions.

NASA   A New Look at the Old Moon
28 minutes. 1980. JSC-749.
The results from Apollo.

EB   The Moon: A Giant Step in Geology
IU   24 minutes. 1976. Rental Fee: EB $48.5; IU $16.50.
Purchase Price: EB $485 (16mm); $390 (video).
Describes Apollo 11 and the analysis of lunar rocks.

NASA   The Time of Apollo
28 minutes. 1975. HQ 229.
A summary of the Apollo project. Breathless prose. Concentration on the contrast between the rusty facilities of the present and the events of the past.

FILMS

The following four AAPT Super 8 films are silent. The Purchase Price of each is $30 U.S./$31 outside the U.S.

AAPT   Lunar Module Descent to the Moon. FN-10.
        The film record of the lunar module's descent to the moon provides a variety of views of the lunar surface from orbit to actual landing.

AAPT   Lunar Samples: Collection and Examination. FN-11.
        Astronauts select and collect a soil sample on the moon. In a laboratory, a technician's examination of some grains of soil under an electron microscope leads to an interesting result.

AAPT   Roving on the Moon (from Apollo 15, 16). FN-18.
        On the Rover with Apollo 15 and 16 crews as they travel to various points on the moon.

AAPT   The Ups and Downs of Being an Astronaut. FN-19.
        The Apollo 16 astronauts jump, trip, stumble, fall, and bounce through their mission near the Descartes Crater.

NASA   The following seven movies are the Apollo Mission Movies. I show a different one each year. The later ones emphasize the scientific aspects more than the earlier ones do.

        Eagle Has Landed: The Flight of Apollo 11
        28 minutes. 1969. HQ194.

        Apollo 12: Pinpoint for Science
        28 minutes. 1969. HQ197.

        Apollo 13: "Houston . . . We've Got a Problem"
        28 minutes. 1970. HQ200.
        The one that didn't make it to the moon. The story of the crisis.

        Apollo 14: Mission to Fra Mauro
        28 minutes. 1971. HQ211. (Spanish version also available.)

        Apollo 15: In the Mountains of the Moon
        28 minutes. 1971. HQ217.

        Apollo 16: Nothing So Hidden
        28 minutes. 1972. HQ222.

        Apollo 17: On the Shoulders of Giants
        28 minutes. 1973. HQ227.

NASA   Mercury/Exploration of a Planet
        24 minutes. 1976. HQa282.
        The flight of Mariner 10 to Venus and Mercury, including an animated sequence on the origin of the solar system.

FILMS

NASA   Venus Pioneer
28 minutes. 1979. NASA #NAV-042.
Post-passage film, including nice animations of atmospheric circulation. Released before radar maps were available, though.

EB   Volcano: Birth of a Mountain
24 minutes. 1977. Rental Fee: $49.50.
Purchase Price: 16 mm $495, video $395.
Fantastic footage of volcanoes erupting in Hawaii Volcanoes National Park.

NASA   Mars--The Search Begins
29 minutes. 1973. HQ236.
An excellent pre-Viking film based on Mariner 9 data. Still worth showing.

NASA   Mars and Beyond
15 minutes. 1976. HQ264.
Pre-Viking. Quick survey and information on experiments.

NASA   Mars: Is There Life?
15 minutes. 1976. HQ263.
Pre-Viking. A lengthy introduction showing a model of Mars' surface, and a brief, reasonable discussion of the biology experiments.

NASA   A Question of Life
29 minutes. 1976. HQ270.
Pre-Viking. Presents an overview of the content of the films "Life?," "Mars: Is There Life?," and "Mars and Beyond."

NASA   19 Minutes to Earth. A Life Science Series movie.
15 minutes. 1978. HQ292.
The results of scientific experiments on Vikings 1 and 2 accompanied by pictures moving distractingly in the third dimension and by jazzy and weird music. 19 minutes is the travel time for radio signals from Mars back to Earth. Conclusions include that liquid water must have flowed, that temperature varies during a day from 190 K to 240 K, that no organic compounds exist; discusses atmospheric composition; states that data from flyby of Phobos may give information on origin and evolution of solar system. Interviews with Charles Hall and Carl Sagan.

NASA   Planet Mars
29 minutes. 1979. HQ283.
A historical introduction (3 min.), Mariner 9 results (3 min.), and then the results of scientific experiments conducted by Vikings 1 and 2.

MLA   Mars: Chemistry Searches for Life
26 minutes. 1978. Rental Fee: $35.
Purchase Price: $494.
> A careful look at the gas exchange on Viking and its results, including slow and detailed explanations of the chemical principles involved. This thorough, one-topic film stands in contrast to the quicker and more superficial treatments of NASA's films.

BU
SU   Mars Minus Myth, revised ed.
CH   22 minutes. 1977. Rental Fee: SU $20 (2-5 days); BU and
IU   KSU (1st ed.) $13 per day; IU $14.65 (3 days).
KSU   Purchase Price: 16mm $400, video $280 from CH.
> Dr. Bruce Murray, a scientist involved in all the major missions to Mars, presents some of the major findings of the Mariner 9 and Viking missions to Mars.

PS   Mars in 3D: Images from the Viking Mission
23 minutes.   1980.   Rental Fee: $25 payable in advance. $100 returnable deposit is required.
Purchase Price: $135.
Written by Elliott C. Levinthal of Stanford University and Kenneth Jones.
> Views in exaggerated 3D of Mars, including views from the orbiter, views of a scale-model of the lander working, and views from the lander. The descriptions are narrated by Dr. Levinthal. While describing how the lander works, the 3D pictures include a view of the lander arm coming out of the screen, the mandatory scene in 3D movies.   The original electronic music is suitable. Special glasses (red and green) are required and are available from PS for .25/pr.; specify number needed when ordering.

Note: Current Jupiter and Saturn films are available on videodisc; a more complete description and ordering information are listed after this section on films.

NASA   Jupiter Odyssey
28 minutes. 1974. HQa243.
> A report on Pioneer 10. Includes interviews with G. Kuiper and G. Munch at the 5-m telescope. Still contains much of interest, even after Voyager.

NASA   Voyager: Jupiter Encounters 1979
8 minutes. 1979. Purchase Price: $85.
> A descriptive sound film about the objectives of the Voyager mission.

AAPT   Voyager Encounter with Jupiter FN-21
4 minutes. 1981. Super-8mm, silent.
Purchase   Price: $30.
> A computer animation showing the Voyager encounter.

FILMS

NASA Jupiter Rotation
2 minutes. 1979. Super-8mm, silent. Purchase Price: $26.
A brief color film showing rotation of Jupiter as actually observed by Voyager 1.

NASA Jupiter's Atmospheric Dynamics
2 minutes. 1979. Purchase Price: $26.
A color film showing the atmospheric circulation observed on Jupiter by Voyager 1 during January 1979. Made from images of Jupiter taken every 10 hours of the same longitudinal region, including the Great Red Spot.

See the Jupiter/Saturn videodisc.

NASA Voyager 1 Encounter Film
3 minutes. 1979. Purchase Price: $30.
A computer-animation film depicting the flyby of Voyager 1 at Jupiter and its Galilean satellites.

T-L Resolution on Saturn
57 minutes. 1982. (WGBH-TV for NOVA) Rental Fee: $85.
Purchase Price: 16mm not available; video $250.
An outstanding film about the Voyager flyby.

IU
KSU Asteroids, Comets, and Meteorites
11 minutes. 1960. Rental fee: $8.70.
A mixture of animation and photos, some of which look outdated. Discusses discovery of and general distribution of asteroids, Halley and his comet, size of comet orbits, head of comet containing particles in gas, direction of comet tails (nice animation), meteorites (nice pictures of laboratory analysis).

WD Comets
11 minutes. 1982. Rental Fee: (16mm only) $35.
Purchase Price: 16mm $299, video $224. Walt Disney Productions.
Animation shows a swirling cloud condensing to form the solar nebula. "How did the solar system begin? The astronomer reaches out to comets with instruments, wanting to know it better." The movie provides a history of the study of comets and describes the ESA Giotto mission, without naming it. The movie is a little too elementary and general, but includes fine animation, as would be expected of Walt Disney Productions.
Walt Disney Productions has several film depositories across the country. For the address nearest you and further ordering information, write to: Walt Disney Educational Media, Marketing Division, 500 Buena Vista St., Santa Barbara, CA 91521 (800) 423-2555.

FILMS

A/CPB   Tales from Other Worlds
FI   60 minutes. 1986. Rental Fee: call 800-LEARNER for closest source; FI (16mm only) $100.
Purchase Price: A/CPB video $198, 3/4" video $298; FI (16mm only) $720.
> The solar system, from the Planet Earth series (a tv series produced by WQED/Pittsburgh in association with the National Academy of Sciences).

NASA   I Will See Such Things...
29 minutes. 1986.
> A description of the Voyager findings about Uranus, narrated by Dr. Albert Hibbs of JPL.
> 3/4" version currently available for loan only from JPL, but will be available from the NASA centers in the future. To order, contact:
> Public Information Office
> Jet Propulsion Laboratory

C/MTI   The Planet That Got Knocked on its Side
58 minutes. 1986. Rental Fee: $125.
Purchase Price: $250 (video only; all formats).
> The Voyager 2 mission to Uranus, narrated in part by the scientists in charge. Solid science from NOVA.
> Transcript available for $4.00 from WGBH Transcripts, P.O. Box 322, Boston, MA 02134 (date shown: 10/21/86).

FILMS

Videodiscs

A breakthrough in presenting slides and films has now been made: all the Jupiter and Saturn Voyager slides and films are available on a single videodisc. Other videodiscs contain other aspects of astronomy or space flight--approximately 100,000 frames each, containing both stills and sound movies. The Center for Aerospace Education at Drew University (CAE) has put together the videodiscs.

The CAE, as of this writing, has made several videodiscs available for $320 to $400. Special educational packages which include the laserdisc player are available. The laserdisc player uses a laser to read out frames, can display single frames with high resolution, and has stereo sound capability. A search function allows you to type in a frame number; the player finds and displays that frame within about 5 seconds. All functions--play, search, etc.--are available on a hand-held remote control, which sends its signals via infrared and so requires no attaching wire.

An interface and "authoring" program have also been develped which allow you to program your own interactive lesons using the laserdiscs and the Apple Computer. The interface works with most laserdisc players.

Space Disc 1 includes:
  Side A: Jupiter
    Summary of all scientific experiments and results
    All stills released by JPL of the Jupiter and Saturn encounters
    Data animation movies of Jupiter and the Great Red Spot
    Plasma wave "music"
    General mission information movie
    Computer-generated mission profile movies of the Voyager
        encounters of the Jovian system, Uranus, and Neptune
  Side B: Saturn
    Summary of all scientific experiments and results
    All stills released by JPL of the Jupiter and Saturn encounters
    Data animation movies of Saturn and ring rotation
    Demonstration of the video-display system that JPL uses to
        construct images in real-time during encounter
    Photopolarimeter stellar occultation movie and data
    Plasma wave "music"
    General mission information movie
    Computer-generated profile movies of the Voyager encounters of
        the Jovian system, Uranus, and Neptune

Space Disc 2: Apollo
Space Disc 3: Space Shuttle
Space Disc 4: The Sun; Side 2 includes the NASA film, UNIVERSE
Space Disc 5: Astronomy
  Includes slides and movies from a variety of observatories.
Space Disc 6: Earth Science
Space Disc 7: Space Age

FILMS

The Space Archive Series ($40 each) includes:

Volume 1: Space Shuttle, Mission Reports STS 5, 6, & 7
Volume 2: Apollo 17 Mission to Taurus Littrow
Volume 3: Shuttle Downlink
Volume 4: Mars and Beyond   (includes 3-D movies of Mars
     as well as movies and stills of the Voyager missions to
     Jupiter, Saturn, Uranus, and Neptune, and of the Galileo
     mission to Jupiter)
Volume 5: Greetings from Earth
Volume 6: Encounters (Voyager 2 flyby of Uranus, coverage of
     Halley's Comet, and physics experiments aboard)

Discs are available from:

Optical Data Corporation
P.O. Box 97
66 Hanover Road
Florham Park, NJ   07932
(800) 524-2481
in New Jersey, call (201) 377-2302

The Instant Replay supplies "Powers of 10" as a special-order
     videodisc; it takes 10-14 weeks to order because it comes from
     Japan. Purchase Price: $65.95 (includes shipping).
To order, contact
     The Instant Replay   VHS-DISC
     479 Winter St.   (800) 847-3472
     Waltham, MA 02154   in Mass.,(617) 890-5384

FILMS

Addresses of Film Sources

A/CPB The Annenberg/CPB collection
1213 Wilmette Ave.
Wilmette, IL 60091
(800) LEARNER; in Illinois, (312) 256-3200

AAPT

American Association of Physics Teachers
5110 Roanoke Place
Suite 101
College Park, MD   20740
(301) 345-4200

BBSO

Big Bear Solar Observatory
Solar Astronomy, 264-33
California Institute of Technology
Pasadena, CA 91125

BFF   Bullfrog Films, Inc.
Oley, PA 19547
(215) 779-8226

BF

Barr Films
PO Box 5667
3490 East Foothill Blvd.
Pasadena, CA   91107

BU

Boston University
Krasker Memorial Film Library
565 Commonwealth Ave.
Boston, MA   02215

CAL

California Institute of Technology
Office of Public Relations, 1-71
1201 East California Blvd.
Pasadena, CA 91125

CH

Churchill Films
662 North Robertson Blvd.
Los Angeles, CA   90069-9990

C/MTI
Coronet/MTI Film & Video
108 Wilmot Road   source for all NOVA films
Deerfield, IL 60015
(800) 621-2131   (312) 940-1260

FILMS

CS

Cinesound
915 North Highland Ave.
Hollywood, CA 90038

EAV

Educational Audio Visual, Inc.
Pleasantville, NY 10570

EB

Encyclopedia Britannica Educational Corp.
425 N. Michigan Ave.
Chicago, IL 60611
(800) 558-6968; in Wisconsin, call collect (414) 351-4488

FI

Films Incorporated/PMI
5547 N. Ravenswood Ave.
Chicago, IL 60640
(800) 323-4222

H-M

Houghton Mifflin Company
For rental or purchase: John T. Fitch, President
                             Kalmia Co., Inc.
                             21 West Circle
                             Concord, MA 01742

ISU

Iowa State University
Film Library, Media Resources Center
121 Pearson Hall
Ames, IA 50011

IU

Indiana University
Audiovisual Center
Bloomington, IN 47405

KAR

Karol Media
22 Riverview Drive
Wayne, NJ 07470-3191
(201) 628-9111

KSU   Kent State University
Audio-Visual Services
Kent, OH 44242
(800) 872-5368
in Ohio, (800) 367-5368

FILMS

MAN

Manitoba Museum of Man and Nature
Audio-Visual Services
190 Rupert Avenue
Winnipeg, Manitoba R3B ON2

MF

Modern Films
PO Box 110100
831 Sable Blvd.
Aurora, CO 80011

MG

Media Guild
11526 Sorrento Valley Road
Suite J
San Diego, CA 92121
619-755-9191 (call collect)

MG-H

McGraw-Hill Training Systems
P.O. Box 641
Del Mar, CA 92014

MLA

Ward's Modern Learning Aids Division
P.O. Box 92912
Rochester, NY 14692-9012
(800) 962-2660   (716) 467-8400
  or, in Canada
c/o Arbor Scientific, Inc.
P.O. Box 113
Port Credit, Ontario

MTPS

Modern Talking Picture Service
5000 Park St., North
St. Petersburg, FL 33709

NASA

See the following section listing NASA film libraries for
your regional distributor.

NCAR

National Center for Atmospheric Research
Media Relations and Information Services Office
P.O. Box 3000
Boulder, CO 80307

NG

National Geographic Society, Educational Services
17th and M Streets, NW
Washington, DC 20036

FILMS

NOAO

Films--Public Information Office
National Optical Astronomy Observatory
Box 26732
Tucson, AZ 85726

NRAO

National Radio Astronomy Observatory
Public Education Office
P.O. Box 2
Green Bank, WV 24944

PBS

PBS Video
1320 Braddock Place
Alexandria, VA 22314
to order (800) 344-3337;   for information (800) 424-7963

PC

Preston Cinema Systems
1621 Stanford St.
Santa Monica, CA 90403
(213) 453-1852

PF

Pyramid Film and Video
Box 1048
Santa Monica, CA 90406-1048

PS

Planetary Society
65 N. Catalina Ave.
Pasadena, CA 91106

SU

Syracuse University
Film Rental Center
1455 East Colvin St.
Syracuse, NY 13210

T-L

Time-Life Video Distribution Center
P.O. Box 644
100 Eisenhower Dr.
Paramus, NJ 07652

UA

University of Arizona
Media and Instructional Services
Film Library
Tucson, AZ 85721

FILMS

WD

Walt Disney Educational Media
Marketing Division
500 Buena Vista St.
Santa Barbara, CA 91521
(800) 423-2355

FILMS

NASA Regional Film Libraries

NASA films, filmstrips and video cassettes can be purchased from:
> The National Audio Visual Center
> National Archives and Records Administration
> Customer Service Section PP
> 8700 Edgeworth Drive
> Capitol Heights, MD 20743-3701
> (800) 638-1300   or (301) 763-1896

(For a list of NASA educational publications, write to NASA Headquarters, Washington, DC 20546.)

All NASA films can be rented free of charge, except for return postage and insurance. To rent films, write to the appropriate address below.

If you live in: Connecticut, Delaware, District of Columbia, Maine, Maryland, Massachusetts, New Hampshire, New Jersey, New York, Pennsylvania, Rhode Island, or Vermont,
> write to: NASA Goddard Space Flight Center
> Public Affairs Office, Code 130
> Greenbelt, MD   20771   (301) 344-8101

If you live in: Alaska*, Arizona, California, Hawaii**, Idaho, Montana, Nevada, Oregon, Utah, Washington, or Wyoming,
> write to: NASA Ames Research Center
> Public Affairs Office, Code 204-12
> Moffett Field, CA   94035   (415) 965-6270

If you live in: Alabama, Arkansas, Iowa, Louisiana, Mississippi, Missouri, or Tennessee,
> write to: NASA George C. Marshall Space Flight Center
> Public Affairs Office, Code CA-20
> Marshall Space Flight Center, AL 35812   (205) 453-4335

If you live in: Florida, Georgia, Puerto Rico, or Virgin Islands,
> write to: NASA John F. Kennedy Space Center
> Public Affairs Office, Code PA-EPS
> Kennedy Space Center, FL   32899   (305) 867-4444

If you live in: Kentucky, North Carolina, South Carolina, Virginia, or West Virginia:
> write to: NASA Langley Research Center
> Technical Library, Mail Stop 185
> Hampton, VA   23665   (804) 865-2634

If you live in: Illinois, Indiana, Michigan, Minnesota, Ohio, or Wisconsin,
> write to: NASA Lewis Research Center
> Film Service Dept.-22

FILMS

21000 Brookpark Rd.
Cleveland, OH 44135   (216) 433-4000 ext. 708

If you live in: Colorado, Kansas, Nebraska, New Mexico, North Dakota, South Dakota, Oklahoma, or Texas,
          write to: NASA Lyndon B. Johnson Space Center
                        Public Information Branch
                        Film Distribution Library, 13
                        Houston, TX 77058
                        (713) 333-4980

*Alaska: selected NASA films are also available from:
                        Alaska State Library
                        Pouch G
                        Juneau, AK 99081

**Hawaii: requests for older NASA films can also be referred to:
                        The Department of Education
                        State of Hawaii
                        Office of Library Services
                        Support Services Branch
                        641 18th Ave.
                        Honolulu, HI 96816

They have not received any new NASA films for the past couple of years.

SOURCES OF AUDIOVISUAL AIDS
(including slides and photographs)

Hansen Planetarium   (wholesale distributor of photographs, slides, and various publications from Palomar, Mt. Wilson and Las Campanas, Kitt Peak, Lick, Anglo-Australian Telescope, etc.)
Space Science Library and Museum
1098 S 200 West
Salt Lake City, UT 84101

Palomar Observatory/Mt. Wilson and Las Campanas Observatories
    catalogue available from:
California Institute of Technology Bookstore
1201 East California Blvd.
Pasadena, CA 91125

Lick Observatory
University of California at Santa Cruz
Santa Cruz, CA 95064

Kitt Peak National Observatory
Public Information Office
P.O. Box 26732
Tucson, AZ 85726

Yerkes Observatory
Williams Bay, WI 53191
    Send self-addressed stamped envelope for listing of publications.

UK Schmidt Telescope Unit
Royal Observatory, Edinburgh
Blackford Hill
Edinburgh EH9 3HJ   Scotland
    A set of 6 Schmidt sky survey transparencies with lab exercises are available for approximately 35 pounds.

AG Astrofotografie
Maxburgsattel
D-6730 Neustadt
West Germany
    Slide sets.

Astronomical Society of the Pacific
1290 24th Ave.
San Francisco, CA   94122
    Planetary and other slide sets.

Astronomy Graphics/Reverse-Image 35mm slides
Dr. Francis S. Lestingi, Dept. of Physics
State University College
1300 Elmwood Ave.
Buffalo, NY 14222-1095
    High-quality graphics to cover an entire course. $0.50/slide after $1.00 each for the first 10.   Catalogue of 445 slides available on request.

AUDIOVISUAL AIDS

CoEvolution Quarterly
P.O. Box 428
Sausalito, CA 94965
   Posters, t-shirts, books, postcards. One poster is of the
   1,000,000 galaxies map.

Cross Educational Software
PO Box 1536
1802 N. Trenton Street
Ruston, LA   71270
   Software for Apple and IBM-PC computers. Volume 10 is Solar
   System Astronomy; Volume 11 is Stellar Astronomy.

David Chandler Co.
P.O. Box 309
La Verne, CA 91750
   Don Davis slide sets, star dials, and other items.

Drift Globe
Star Route Box 38
Winthrop, WA 98862
   A 12" globe with continents attached by velcro; continents can be
   re-arranged according to marked positions to illustrate plate
   tectonics. $149.50 postpaid.

Everything in the Universe
Norm Sperling
5248 Lawton Ave.
Oakland, CA 94618
   Norm Sperling has made available a variety of astronomical
   objects, including Owen Gingerich's set of slides from historical
   books and a miscellany of telescopes, books, posters, and other
   astronomical objects. Write for catalogue.

Holiday Films
Whittier, CA 90608   (213) 945-3325
   Films and slides.

Hubbard
P.O. Box 104
Northbrook, IL 60062
   The following 8mm astronomy film loops are available:
   9100   Set of 6 with lesson plans $162.
   9101   Mo   9102   planetary Motion
   9103   Earth: Rotation and Revolution
   9104   Seasons
   9105   Time and Dateline
   9106   Day and Night
                   Individual Titles $29.50

   In addition, the following 35mm slides are available:
   286   Galaxies--Nebulae--Star Clusters
                   15 color slides $21.50
   287   Solar Phenomena
                   15 black-and-white slides $21.50

288   Moon--Planets--Comets
            15 black-and-white, 3 color slides $21.50
296   The Lunar Surface
            20 color slides $25.75

Galaxy Productions
15522 Vanowen St.
Suite 104
Van Nuys, CA 91406
  Posters, slides, movies, and tapes.

KaiDib Films
P.O. Box 261
Glendale, CA 91208-0261
  Slide sets include "Planets Close-Up," a set of 160 slides for
  $210.00 produced through special arrangement with JPL and NASA.

MMI Corporation
2950 Wyman Parkway
Baltimore, MD 21211
  Astronomy Materials catalogue, including portable school
  planetariums, available free to educators writing on school
  letterhead (others: $2).

David New
P.O. Box 278
Anacortes, WA 98221
  Sells specimens of meteorites and tektites; write for current
  specimen list.

Optica b/c
4100 MacArthur Blvd.
Oakland, CA 94619
(415) 530-1234
  Publications & Audio-Visual Aids Catalogue. Astrophotography
  Systems Catalogue. Astronomical Events Calendar free with
  self-addressed stamped envelope.

Photographic Illustration Co.
P.O. Box 6699
Burbank, CA 91510
  Slides and prints of NASA photos of Saturn and its moons.

Science Graphics, Inc.
P.O. Box 17871
Tucson, AZ 85731
  Catalog includes: Solar System and Stellar Systems Astronomy
  (650 slides), Physical Geology (720 slides), Planetary Geology
  (200 slides). Catalog free to instructors writing on letterhead
  (others: $1.50). New Product newsletter sent free. Science
  Graphics is also now a distributor of all 35mm astronomical
  slides released by the National Optical Astronomy Observatories,
  including Kitt Peak, Cerro Tololo, and Sacramento Peak Solar
  Observatory.

AUDIOVISUAL AIDS

Sky Publishing Co.
49 Bay State Rd.
Cambridge, MA 02238
   The publishers of Sky and Telescope magazine. They distribute
   high-quality photographic prints of astronomical objects.

Tersch Enterprises
P.O. Box 1059
Colorado Springs, CO 80901
(303) 597-3603
   Write for a free catalogue listing over 3,500 slides on
   astronomical and spaceflight topics. A slide set with 90 color
   and black-and-white slides costs $65.

Woodstock Products
P.O. Box 4087
Beverly Hills, CA 90213
   Brochure on color space photographs.

AUDIOVISUAL AIDS

In addition to publishing astronomy journals and magazines, the following organizations are also sources of audiovisual materials:

American Association of Physics Teachers
Executive Offices
5110 Roanoke Place
Suite 101
College Park, MD   20740
 AAPT Products--a catalogue of physics and astronomy films, slides, and publications.
 Also available from the AAPT is Women in Science, by Dinah L. Moche, a series of 6 interviews ranging from 11 to 18 minutes each. Purchase Price: $30 ($32.50 outside U.S.); payment must accompany order. Includes audio cassette, slides, and articles about 6 women with careers in science and engineering.
 The scientists interviewed in this series are:
  Virginia Trimble, Associate Professor of Physics at the University of California, Irvine, and Associate Professor of Physics and Astronomy at the University of Maryland, College Park;
  Diana McSherry, a research bio-physicist and Executive Vice President of Digisonics Corporation;
  Gwendolyn Albert, an environmental engineer with the Southwestern Division of the U.S. Army Corps of Engineers;
  Chien Shiung Wu, Professor of Physics at Columbia University;
  Betsy Ancker-Johnson, Associate Director of Argonne National Laboratory in charge of physical research;
  Carolyn Leach, Chief of the Biomedical Laboratories Branch at the NASA-Johnson Space Center, Houston, Texas.

American Institute of Physics
MD 12
Marketing Services
335 East 45th Street
New York, NY 10017
 Newly available: Moments of Discovery, comprising two audiovisual units. Unit 1 describes the discovery of fission. Unit 2, A Pulsar Discovery, includes the (accidental!) live recording of the optical pulsing of the Crab pulsar. Both units contain audio cassettes with the voices of many scientists, slides (18 on fission and 2 on the pulsar), 2 illustrated student scripts, and teachers' guides. Purchase price: two units, $85; extra set of 10 scripts for either unit, $10.

AUDIOVISUAL AIDS

Astronomical Society of the Pacific
1290 24th Ave.
San Francisco, CA 94122
   "The Solar System Close-up," two sets of 50 slides each with
captions assembled by Dr. David Morrison. Describes the results of
planetary exploration during the past 2 decades. Price: $34.95 +
$4.50 postage and handling per set, or both sets at $74.95 (includes
postage and handling).
   Newly available: a third set of 30 slides in the solar system
series, with a 60-minute audiotape narration. Price: $24.95
(includes postage and handling).

   Publishes Astronomical Products Catalog--a catalogue of astronomy
photography, posters, slides, star maps, astronomical software,
cassettes about astronomy and observing, bumper stickers ("Black
Holes Are Out of Sight" and 9 others), t-shirts, and reprints from
Mercury magazine.

   Also produces a series of information packets on astronomical
topics including "Selecting Your First Telescope," "Astronomy and
Astrology," "The Return of Comet Halley," "Quasars," etc.

Astronomy Slides
Astromedia Corporation
P.O. Box 92788
Milwaukee, WI 53202
   Set of 41 space art slides, available either individually or in
sets. Star maps, posters, and other items are also available. See
Astronomy magazine for the latest ads.

Sky and Telescope
49 Bay State Rd.
Cambridge, MA 02238
   Check the many advertisements in this magazine for the latest items
and for additional sources. They also distribute many books and
star atlases (including Wil Tirion's magnificent Star Atlas 2000.0).

AUDIOVISUAL AIDS

Star Maps

David Chandler Co.
P.O. Box 309
LaVerne, CA 91750

    1) A planisphere, a star map on a wheel that rotates within a mask so that it can be set to match the sky for any day and time. Die-cut for latitude zones 20°-32°, 30°-40°, or 38°-50° and 30°-40° S for Australia and New Zealand. Price: $3.25 each + postage or a minimum quantity of 25 for $1.95 each.

    2) Deep Space 3-D: A Stereo Atlas of the Stars. The sky is covered on 14 5" x 8" cards, each having a reference chart and a pair of computer-generated stereo views that amplify the parallax. A folding stereo viewer is included. Price: $7.95 each + postage or a minumum quantity of 10 for $4.75 each.

    3) Halley's Comet model: Three-dimensional model of the comet's path relative to the sun and earth during its 1985-86 appearance. Price: $1.25 each + postage or a minimum of 10 for $1.00 each.

Games

Sky Challenger
Discovery Corner
Lawrence Hall of Science
University of California
Berkeley, CA 94720

    Star-finder kit with 6 sky wheels for locating different groups of constellations and other objects. Includes a guide with suggestions for using the kit. $5.95 + $2.50 shipping for one. California orders must include correct sales tax.

"Stellar 28" Constellation Games

    A kit for learning to locate and identify stars and constellations. Includes game board, cards, and instructions. Developed by Neil Comins of the University of Maine and Ronald Littlefield of NASA's Goddard Research Center. Available from Hubbard Scientific Company, P.O. Box 104, Northbrook, IL 60062.

COORDINATION

American Association of Physics Teachers

Resource Letter EMAA-1:   Educational Materials in Astronomy and Astrophysics. AJP 41, June 1973, pp. 783-808.
Resource Letter EMAA-2:   Laboratory Experiences for Elementary Astronomy. AJP 44, September 1976, pp. 823-833.

The chapters of EMAA-1, by Haym Kruglak, include the following topics: primary sources, slides and prints, films, film loops and audiotapes, books, yearbooks and compendia, ephemerides and atlases, laboratory exercises, term paper topics, and equipment.   The 1976 update includes laboratory manuals, articles, sources of exercises, planetariums, films and film loops, telescopes, and distributors.

Reprints are available.   Request Resource Letter EMAA-1 and EMAA-2.   Enclose $1 per copy (not in stamps) of EMAA-1, $.75 per copy of EMAA-2, and a self-addressed, stamped envelope to:
> Executive Office
> American Association of Physics Teachers
> 5110 Roanoke Place
> Suite 101
> College Park, MD   20740

The AAPT also publishes the American Journal of Physics and The Physics Teacher, both of which have articles and reviews of interest to teachers of astronomy.   Also available are books of scientific reprints, such as Cosmology, a collection of sixteen articles on research in cosmology (L.C. Shepley and A.A. Strassenburg, eds. 1979, 212 pages, $6).

The AAPT's Astronomy Education Committee is working to enlarge the role of astronomy in the AAPT.

* * * * *

American Astronomical Society
Education Office (Charles R. Tolbert, Education Officer)
Leander McCormick Observatory
University of Virginia
PO Box 3818, University Station
Charlottesville, VA 22903-0818

The AAS has a booklet on Careers in Astronomy ($0.25 requested). The AAS's Harlow Shapley Visiting Lectureships Program sponsors and subsidizes two-day visits by astronomers to junior colleges, colleges, and universities in the U.S. and Canada, especially to those that have few or no professional astronomers teaching.   Over 100 visits are made each year, typically including a public talk, a guest appearance in a class, a seminar for the department, and private discussions with faculty and deans about improving astronomy content in the curriculum. For information contact Dr. Tolbert.
The AAS has also set up a Task Group on Education in Astronomy (TGEA) to coordinate materials relevant to teaching.   All teachers of astronomy should get on the mailing list by writing to the AAS Education Office.

Materials that have been available from time to time include lists of laboratories, syllabi, and audiovisual material. A workshop on teaching has also been given.

* * * * *

1) An Astronomy Materials Resource Guide, 191 pages, updated as of 1985, is available. Send $7.50 ($6.25 + 1.25 postage and handling--$0.50 extra shipping for second and succeeding items) to:

> Astronomy Resource Guide
> West Virginia University Bookstores
> College Avenue
> Morgantown, WV 26506-6122

For inclusion of materials (unpublished) to the next updated guide, write:

> Dennis W. Sunal
> Astronomy Education Materials Network
> 604 Allen Hall
> West Virginia University
> Morgantown, WV 26506

2) A laboratory guide for secondary and undergraduate courses titled Model of Nearby Space is available. Used by students independently or as a class, the interconnected series of exercises helps the students make simple angular measurements which finally result in a paper 3-D construction of the solar system and nearby stars (to about 500 light years). The appendix has outlines of planetarium programs (also available) and student worksheets.

The Model of Nearby Space, 70 pages, is available for $6.25 plus $2 postage and handling from the bookstore above ($.50 extra shipping for second and succeeding items).

* * * * *

The National Science Teachers Association
1742 Connecticut Ave., NW
Washington, DC 20009

The NSTA is a professional organization devoted to the improvement of science teaching at all levels. It sponsors local, national, and international workshops and conferences, and provides special publications at low cost on topics of interest to its members. The NSTA also publishes a news bulletin and three journals devoted to science teaching at different educational levels.

The Journal of College Science Teaching is devoted to science teaching at the college level, principally in introductory courses and in courses for non-majors. It includes articles by distinguished science educators; "How We Do It" articles concerned with innovative courses, techniques and materials; and regular departments covering books, instructional media, legislative news, science news, and analyses of research.

> Membership rates:
> College: $30 (includes subscription to the Journal of College Science Teaching)
> Comprehensive: $55 (includes subscriptions to any two journals plus 2 or 3 special publications)
> Comprehensive Plus: $65 (includes subscriptions to all three journals plus 2 or 3 special publications)
> Student: $20 (includes subscription to one magazine)

* * * * *

Astronomical Society of the Pacific
1290 24th Ave.
San Francisco, CA   94122

Publishes 2 journals, regional directories of astronomy activities, monthly sky calendars and star charts, introductory information packets, bibliographic materials, and provides other educational services and activities.
Also publishes a free index of NASA astronomy books, an Astronomical Products Catalog (see section on audiovisual aids, p. 47).
Annual membership is $21; for information, write to the Membership Dept. and enclose a legal-sized self-addressed stamped envelope.

* * * * *

The Planetary Society
65 N. Catalina Ave.
Pasadena, CA   91106

The Planetary Society is a non-profit organization of both scientists and non-scientists devoted to supporting and participating in the exploration of the solar system, the search for planets around other stars, and the search for extraterrestrial life. The Society publishes The Planetary Report, a members-only newsletter, and sponsors numerous conferences, seminars, workshops, and other events. Books, photographs, slide sets, posters, and other materials are offered at discount prices through the newsletter.
Membership: $20 annually, $25 for foreign.

* * * * *

The Viking Fund
American Astronautical Society
P.O. Box 7205
Menlo Park, CA
The Viking Fund has been set up by the American Astronautical Society to collect donations to give to NASA. All contributions are welcomed and may be sent to their address above.

* * * * *

ORGANIZATIONS AND PUBLICATIONS

National Space Institute
West Wing, Suite 203
600 Maryland Ave., S.W.
Washington, DC   20024
(202) 484-1111

The NSI publishes an interesting magazine and newsletter, and is active in supporting space research and in lobbying.

* * * * *

The International Astronomical Union has a commission on the teaching of astronomy. I am the U.S. national representative. Please notify me of any information that you would like transmitted to the international community of astronomy teachers.

* * * * *

NASA Report to Educators.

Published four times a year. Write to:
NASA
Educational Programs Division
Office of Public Affairs, Code FE
Washington, DC 20546

NASA publishes a variety of educational materials that describe and explain NASA aerospace programs and projects, advances in science and technology resulting from these programs and projects, benefits on Earth of aerospace technology. These publications include pamphlets, bibliographies, facts sheets, posters, and picture sets. All NASA publications and curriculum resources materials should be ordered from:
Superintendent of Documents
U. S. Government Printing Office
Washington, DC 20402

NASA photographs (either 4" x 5" transparencies or black-and-white 8" x 10" glossies) are also available. For ordering information and an index of NASA photographs, write to:
Audiovisual Section
LFB-10/Office of Public Affairs
NASA Headquarters
Washington, DC 20546

REVIEWS OF RELEVANT BOOKS AND FILMS

Sky and Telescope
49 Bay State Rd.
Cambridge, MA 02238
      Subscription price: $15 per year. (Canada and Mexico add $3.50
      per year.)

Astronomy
Astromedia Corp.
411 E. Mason St.
P.O. Box 92788
Milwaukee, WI 53202
      Subscription price: $21 per year. (2 yrs. $39, 3 yrs. $55)
      Foreign: add $5 per year.)

      Odyssey, a children's astronomy magazine, is also published by
      Astromedia. Subscription price: $16 per year. (2 yrs. $26)

AAAS Science Books and Films
Subscription Dept./10th Floor
1101 Vermont Ave. N.W.
Washington, DC 20005
      Subscription price: $20 per year, $18 per year for AAAS members.

      Films in the Sciences
        Michele M. Newman and Madelyn A. McRae, eds., 186 pages.
      Guide to nearly 1000 of the best science and technology films
      produced from 1974 to 1980, arranged by subject. Contains
      critical reviews. Order from the AAAS Sales Dept., $14/copy.
      (AAAS members deduct 10% on prepaid orders.)

American Journal of Physics and The Physics Teacher
American Association of Physics Teachers
5110 Roanoke Place, Suite 101
College Park, MD 20740
 For editorial correspondence:
  AJP: John S. Rigden, Editor
  240 Benton Hall
  University of Missouri at St. Louis
  St. Louis, MO 63121
  TPT: Donald Kirwan, Editor
  P.O. Box 336
  Kingston, RI 02881
      Published by the American Association of Physics Teachers.
      Regular membership in the AAPT is $29 and includes a subscription
      to one journal, with a special price for a second.

Mercury
Astronomical Society of the Pacific
1290 24th Ave.
San Francisco, CA 94122
      A bi-monthly magazine published by the ASP. Membership in the
      ASP is $21 per year ($28 foreign) and includes a subscription to
      Mercury.

OBSERVING GUIDES

The Astronomical Ephemeris
    Available over a year in advance from:
        Superintendent of Documents
        U.S. Government Printing Office
        Washington, DC   20402

The Explanatory Supplement to the Astronomical Ephemeris and The
    American Ephemeris and Nautical Almanac
    Available from:
        Pendragon House, Inc.
        2898 Joseph Ave.
        Campbell, CA   95008

A Field Guide to the Stars and Planets, second edition, Donald H.
    Menzel and Jay M. Pasachoff, Peterson Field Guide Series,
    Houghton Mifflin, 1983.
    A completely new book, including monthly sky maps, a full Tirion
    Star Atlas, Graphic Time Tables to locate bright stars, double
    and variable stars, nebulae, galaxies, and planets, color plates,
    and descriptive text.

Floppy Almanac
    A new compact source of accurate astronomical information is
    being developed at the U.S. Naval Observatory. A 5-1/4 inch
    computer diskette will contain the principle data now contained
    in three annual Observatory publications--the Astronomical
    Almanac, the Nautical Almanac, and the Air Almanac. For
    IBM-compatible computers. Projected price: $20, available late
    1986.
    Contact:
        Dr. George H. Kaplan   (202) 653-1516
        Nautical Almanac Office
        U.S. Naval Observatory
        Washington, DC 20390

Skywatcher's Alamanac
    Yearly publication with a computer-generated local sun and moon
    calendar with information for specified latitude, longitude and
    time zone. Other publications include Local Planetary Visibility
    Report, Local Sidereal Time Tables, and Daily Sun Angle Report.
    Available from:
        Astronomical Data Service
        3922 Leisure Lane
        Colorado Springs, CO 80917

LAB MANUALS

Blitz, Leo. Experiments in Astronomy, Burgess Publishing Co., (1983).

Christiansen, Wayne A., Ronald H. Kaitchuck, and Michelle D. Kaitchuck. Investigations in Observational Astronomy, 2nd edition, Paladin House (Geneva, IL: 1978).

Culver, Roger B. An Introduction to Experimental Astronomy, Freeman (New York: 1984).

Paul E. Johnson and Ronald W. Canterna. Laboratory Concepts for Introductory Astronomy, Saunders College Publishing (Philadelphia: 1987).

Kelsey, Linda J., Darrel B. Hoff, and John S. Neff. Astronomy: Activities and Experiments, 2nd edition, Kendall/Hunt Publishing Co. (Dubuque, Iowa: 1983).

Safko, John L.. Laboratory Exercises in Astronomy, self-published, 1974.

Tattersfield, D. Projects and Demonstrations in Astronomy. John Wiley & Sons (New York: 1979).

TRANSCRIPTS AND NOTES ON MOVIES

Students enjoy movies, but to ensure that they remember what was said and shown, it is helpful to distribute a transcript of the movie. The following is a transcript of the movie, Birth and Death of a Star--one of the best astronomical movies.

(John Wheeler, Narrator): From the beginning of recorded time, the same star, the sun, has been observed in the sky burning with the same apparent brightness, giving off familiar amounts of life-sustaining heat and energy. Further out in space the universe is populated by billions of other stars that never seem to change. Generations of star watchers, philosophers, thinkers, and others have passed into history; stars have remained apparently unchanged. No one has witnessed stellar birth or death, and the idea that the sun and all the stars have been here from everlasting to everlasting was a very old idea, but it was a wrong idea. And the idea that the stars, like animals, have a birth, a main life, and a death has a recent origin.

And think of what it is not just to tell what the stars are, and where they are, but to tell also how that star got there, and what it was like 10,000 years ago, and what it was like 3 billion years ago. It's a story, a history book, a novel, that we're trying to open up and read. We're having to take information from a variety of sources and put it together to make a story, but the marvelous thing is that we don't give up hope on the story--that we give it a try.

To unfold this story we have scientists with theories about stars, who make observations with large radio and optical telescopes. Perhaps somewhere in the search among the stars a door will open up, and behind that door will be revealed all the glittering central machinery of the universe, in all its simplicity and beauty.

The story begins in a vast cloud of gas and dust made up primarily of the simplest element--hydrogen, a few heavier elements and a general debris of space; but the content of the cloud is changing, and a building process is taking place. Elements seem to naturally combine with each other to form molecules. These molecules have been detected by radio telescopes, like the radio telescope of the National Radio Astronomy Observatory on Kitt Peak near Tucson, Arizona.

John Ball of Harvard University is one of a group of electrical engineers, chemists, and radio astronomers who have discovered a number of these molecules in interstellar space.

(John Ball): We identify a molecule by the particular frequency it radiates in much the same way that a radio station is identified by its frequency on a dial. Actually, the molecules we are looking for radiate a whole series of different frequencies. Over two dozen molecules, some with as many as 7 atoms, have been discovered by radio techniques in interstellar space. Evidently there is a common widespread tendency toward building or bringing together elements to form complex structures. How complicated the molecules can become in interstellar space, we're not sure. Other processes are needed to build stars, planets, and living things.

(Narrator): Clues to the nature of these other processes have been found by comparing a variety of clouds. One common feature is a number of much smaller dark clouds that dot the surfaces of many of the larger clouds. On the same mountaintop in Arizona, using an optical rather than a radio telescope, Beverly Lynds of the Kitt Peak National Observatory has been studying and cataloging the dark clouds.

(Beverly Lynds): These are fascinating regions of the sky which stand out because of the absence of stars. Very little information was available about these regions when I became interested in them. I am using the telescope as a large camera to photograph the dark clouds. Now that may seem strange--to try to photograph things that don't give off light; but the dark clouds can be photographed, because they are outlined in starlight. We have to be rather ingenious to develop ways in which we can find out more details, more of the vital statistics of the dark clouds. Where are they? How many of them inhabit space? What are they made of? Using the telescope and the camera is one way of studying them. Photographs show a very special class of blue stars in the clouds, stars we believe to be the youngest in the galaxy. They are so young that they haven't had time to fly from their nest. So it appears that the nest is the dark clouds, and this is where these young stars are born. One of the most interesting features is that as these clouds get smaller, they get darker. This rather suggests that perhaps they began as great tenuous clouds which, as they collapse, get darker, and denser, and blacker. But we can't see inside the nest. No one has witnessed stellar birth. Theoreticians have written the story of how it probably occurs. . . .

In theory, when a cloud of gas and dust reaches a certain density, it could collapse or fall together under its own weight. It might fall into one or more separate units. The larger ones could become stars; the smaller, planets. The volume shrinks a billion billion times. The cloud becomes hotter and denser as it continues to fall in toward itself. At the center of the larger clouds the densities and temperatures might be so high that there are collisions between the atoms of hydrogen, and some of these collisions are energetic enough to produce nuclear reactions. The release of energy creates a gas pressure that stops the collapse of the clouds. Some clouds are unstable and move in and out, or pulsate, as the forces of gas pressure and gravity seek to adjust themselves. But eventually, after millions of years, as the hydrogen burns in the furnace of the interior, a stable cloud of gas--a stable star--one of nature's most magnificent inventions, is born.

(Narrator): The star might remain stable for billions of years until it reaches middle age. The sun is one of countless numbers of stable middle-aged stars. Making observations during the day rather than at night on the same mountain high above the Arizona desert, Don Hall of the Kitt Peak National Observatory has been exploring the surface of the sun.

(Don Hall): Because the sun is so close and bright we have to use a different kind of telescope than we use for the distant stars. The solar image is much too hot to view directly. Instead the image is reflected by a number of mirrors onto a screen, in the same way that movies are projected on a screen. The image can then be viewed, photographed, or sent on to various instruments. One can see details on the surface, which is not possible to see on any other star. We can study the motion of these details with photographs taken over a period of time. Like all stars, the sun turns about its axis. Its rotation period is about 28 days, as we measure from seeing features like sunspots and hot areas drift across the surface. The surface of the sun is a constantly turbulent boiling gas, at a temperature as hot as 6,000°. There are boiling cells 500 or 600 miles across. On a larger scale one sees prominences, which are long arches of gas, flowing out along the magnetic fields for tens and even hundreds of thousands of miles. Above the surface, the solar atmosphere heats up to about a million degrees--instead of getting cooler as you might expect. We still don't have a very good idea of how these layers are heated, or what creates them, or keeps them in place. Another mystery is the way in which flares happen. These are vast explosions that give off great streams of x-rays and cosmic rays, which eventually reach the earth, and in some cases disturb radio communications. We still don't really understand what the source of energy is, or the mechanism by which things happen during the flare.

The instrument here is breaking the light up into its colors or wavelengths, very much in the same way as this diffraction grating breaks the light up into colors--only the instrument here is very much bigger and more powerful, and it allows one to break the light up into very fine wavelength intervals, and then to search these for the presence of chemical elements, compounds. Elements that I've discovered include hydrochloric acid; hydrofluoric acid, which was the first time we knew that fluorine was in the sun as well as in toothpaste; water vapor; and mainly isotopes of various elements. This work may eventually lead to new theories of the way in which the sun was born-- the way in which stars develop. There are lots of other stars like the sun, and there are probably many stars like the sun which have planets around them--even planets like the earth, which may be capable of supporting life. We will have a few billion years to search for these, because current theories indicate that the sun will remain in its present stable state for at least this long. At some distant time, however, the sun will run out of nuclear fuel, and then catastrophic changes will begin to take place.

(Jesse L. Greenstein): The sun becomes a red giant. It bloats up in the sky. It becomes 100 times brighter than it now is. The oceans melt. The atmospheres have evaporated. This giant red blob grows bigger and bigger in the sky, as it eventually . . . eventually the outer atmosphere of this giant red star evaporates into space. And left shining is a little, brilliant, hot ball--too small to distinguish from the other stars--the beginning of the white dwarf stage.

(Narrator): On Palomar Mountain in southern California, hidden from the lights of the city of Los Angeles, is one of the most famous optical instruments in the world--the 200-inch Hale Telescope. Jesse L. Greenstein of the California Institute of Technology photographs and studies the images of faint and aging stars.

(Greenstein): I'm working on a class of stars called white dwarfs.

They are tiny little things, about the size of the earth, only they weigh as much as the sun. I have to use the big telescope, just because there is too little light. The area of the telescope is about 600,000 times the area of the eye at night, and we need that much extra to capture the faint glow of the star. Without this monster, this dream of 500 tons of glass and steel, we couldn't hope to find out as much as we do. Not an awful lot of information—enough to tell us what the star is made of, how hot it is, how powerful the forces of gravity are, whether our theories are correct. What is it? A little spot of light? I can barely see it. Signals from a dying star. Pretty near us. The star I am working on is probably ten billion years old. For five billion years it was a star like the sun—brilliant, filled the sky of its planets if it had planets. Now it is a little dying dot no larger than the earth with the matter compressed by gravity to densities such that in the center it weighs nearly 100 tons per cubic inch. But no one on earth can imagine what a solid is like that weighs nearly 100 tons per cubic inch and has a temperature of 30 million degrees. We are trying to see what it is made of, how great the pressures are, where it is from now. It'll be our sun. An old man's occupation, watching it happen. Eye of the mind, ten billion years. It's a kind of philosophic calming process, maybe, to sit here at night, and watch this faint glow, and think of the time involved, and think of the long and exciting processes that led this way. And you know you have one ultimate limit, far beyond that of the white dwarfs—an extraordinary thing called the neutron stars.

(Frank Drake): Theoreticians predicted that when a very massive star collapsed, it might fall right through the white dwarf stage. Then it might explode in one of the greatest displays of fireworks in the universe.

The collapsing star, or the part left behind by the explosion, might continue to collapse until the elementary particles, such as neutrons, inside the atoms could be compressed no further. This theory of the formation of a new kind of star called the neutron star was kicking around for many years. During this time radio astronomers were tuned in to stars and galaxies which emitted only a steady hissing that never seemed to change year after year.

(Narrator): The Arecibo radio telescope in Puerto Rico has a larger surface than any other telescope in the world. Frank Drake of Cornell University was one of the first to use this enormous instrument.

(Drake): And then one day we received a report that there was a cosmic radio source, very far from the earth, which was turning on and off in thousandths of a second. It was an almost preposterous thing—I thought that it was a joke. We came to this telescope here, and pointed in the direction of this recording that I thought was a hoax. And to our amazement the chart recorder, the first time I'd ever seen such a thing happen, started going wildly back and forth, about every second. The intermittent radio source, which we now call a pulsar, really existed! It was different from anything we had ever seen. It was an absolutely amazing sight to see.

Of course, part of the excitement was that the signals from the pulsar resembled so closely the radiations we'd expect to receive from other intelligent civilizations. They were pretty strong, but it wasn't out of the question that that is what they were. And, of course, all the early observers--including myself--looked very closely at the record to see if we could see anything in them that would show that they were really intelligent signals after all--which would, of course, make them probably the greatest discovery of all time.

To make such measurements in such a short time requires us to gather a lot of energy in that short time. Radiation from the star is focused by the enormous reflector, or dish, to a point--in fact, a line five hundred feet in the air--where the information is fed to a receiver, and then to the chart recorder or computers. One pulsar was found to have its period of pulsation getting slowly longer by 36 billionths of a second a day. That doesn't sound like much, but it was actually a crucial and critical discovery. In one fell swoop, in one instant, it showed us that not only were the pulsars natural sources, but they were, in fact, spinning tops, gradually slowing down. As the massive star collapses, the rate of spin is very greatly increased and intense magnetic fields are built up. The result, a rotating beacon, which sweeps the cosmos. If it happens to point in our direction as it sweeps, we observe flashes of energy. Eventually the spinning slows down as the energy is used up--but that isn't the end of the story.

(Narrator): I suppose it is just unbelievable to think that there is something still more compact, still denser, still more mysterious, than a neutron star. But so it is: the black hole. Although we haven't seen one yet, we feel that the discovery of the first black hole is around the corner--probably the evidence is staring at us right now.

This star, if you think about it, has got a certain amount of matter. And if there is too much matter there, there's just absolutely no way to support it. It'll just keep on going in, and collapsing. And this collapse will end up with a stage that makes one think of nothing so much as I can think of as the story of the Cheshire Cat in Alice in Wonderland, that faded away on the branch of the tree with nothing left behind except the grin. When matter collapses completely to a black hole, all our evidence goes to tell us that what we have left behind is only gravitational attraction. A planet going around will still go around. Attraction is still there, but there is no light coming out--no radiation trying to escape from that object can get away from it. And, in this sense, it is completely black. Anything you drop in is lost--all its features disappear.

I hope I can give a little bit of a feel for what a black hole might look like. I suppose there is no better way to do it than to imagine oneself on a trip to a black hole in a space capsule. There is a whole area of the sky ahead where one can see no stars. As we come up to the black hole, and we take a turn around it, we're under its gravitational pull all the time. But what do we see? Well, we see the sky and the stars sweeping around, as if we were in an aeroplane taking a turn. But then, next, no black hole is ever going to be sitting all by itself. There is going to be matter around it--

falling into it. Drop in stones, or bricks, or atoms, dust, clouds, solids and liquids, and they all end up in this object--with no observable properties, no light coming off, nothing but simple gravitational attraction. As the matter streams in, it builds up a traffic jam like the cars streaming in from all directions to a baseball game, their red lights blinking. In one way it is preposterously small compared to the original star. But in another way, it is preposterously large--because the black hole is 1, or 2, or 3 miles across, or even more. Of all the strange forms of matter, of all the strange things that can happen on heaven or on earth, the strangest is this collapse--and the most exciting to look for. And you might say, "Well, what reason have we to believe that such an object can exist?" And the answer is: We see absolutely no way to escape the existence of such objects. And then you can say, "Well, if a black hole is black--how are you ever going to see it?" And surely that last place in the world to look is to hope to see a black hole by looking directly at it. The real hope to see a black hole is in conjunction with a normal star. We know that many more than 10 percent, perhaps more than a quarter of all the stars in the universe, are double stars. So we expect to see black holes in conjunction with an ordinary star. But why should one be so interested in seeing one of these objects? What good is the black hole, anyway? And what we can say, really, is that looked at in the biggest sense of the word, a black hole is a kind of . . . well, I would call it an experimental model, if you will, for the collapse of the universe that Einstein tells us should happen at a later date, in the final stages of our universe.

But, also, a black hole is of interest in a more immediate way in filling out this novel that we're talking of. We won't feel that this story is really done until this novel is opened up, and in the novel we see page after page of these excitements--these characters in the history of our universe. The young blue giants. The middle-aged, middle-sized average star, which we know can sustain life. The star that swells to a bloated all-consuming red giant. The white dwarf. And the neutron star. This last: the black hole. We say--here is where our real hope is of seeing what this universe is, and how we fit into it.

Measured on the time scale of human history, stars seem to be everlasting. But on the cosmic time scale billions of years, they are born, live out their lives, and disappear from view. Then, in darkness, the remnants of exploded and burned out stars become part of the debris of space--debris that goes into the making of new stars. Second, then third generations, slowly evolve to some unknown and distant fate. Birth, life, death seem to be universal.

****

This transcript was reprinted with the permission of the American Institute of Physics.

TRANSCRIPTS AND NOTES ON MOVIES

The following are notes from the movie Crab Nebula:

Chinese chronicles recorded a "guest star" in 1054 A. D.

We now study the Crab Nebula with optical telescopes, radio telescopes, balloons, etc.

Philip Morrison (Massachusetts Institute of Technology): The Crab Nebula is a Rosetta Stone to understand the life and death of stars, with the difference that for the Crab Nebula we don't know any of the languages in which it was written.

We can see actual changes in the Crab Nebula: we can see it growing --expanding at 1300 km/sec (800 miles/sec).

Circumstantial evidence links the Crab Nebula and the Supernova of 1054. Some controversial evidence from Indian drawings in the American West exists. Why is there no record of European sightings? After all, from the comet displayed in the Bayeux tapestry, we know they watched celestial events.

The Earl of Rosse gave the Crab Nebula its name; he observed it with his mammoth 180-cm (72-in) telescope a century ago. It was already M1 in the catalogue that Messier had compiled of fuzzy objects that comet searchers might confuse with new comets.

Joseph S. Miller (Lick Observatory): Color is visible only on long exposures. Most of the stars on the image don't have anything to do with the nebula. An amorphous region and the filaments do. The filaments represent material enriched in heavy elements (for example: carbon, oxygen, and iron) inside a star, and that material is now spreading into interstellar space.

F. Graham Smith (Columbia University): The Crab Nebula was one of the first radio sources detected. A clear picture required interferometers--several telescopes linked. What keeps the Taurus A radio source--the Crab Nebula--hot? Its light is polarized; ordinary glowing gas is not polarized, but synchrotron radiation is polarized.

Malvin A. Ruderman (Columbia University): Neutron stars--surface could support foot-high mountains; atmosphere a few inches high.

At Mullard Radio Observatory in Cambridge, England, 4.5 acres of aerials.

S. Jocelyn Bell discovered pulsing source that followed sidereal time (time by the stars). Anthony Hewish, the head of the group, is also shown.

Crab pulsar discovered with the 91-m (300-ft) telescope at National Radio Astronomy Observatory in Green Bank, West Virginia. David H. Staelin (MIT) working with Edward C. Reifenstein found the Crab pulsar from its dispersion.

TRANSCRIPTS AND NOTES ON MOVIES

The pulsar's period was discovered with the 304-m (1000-ft) telescope at Arecibo, Puerto Rico. Described by <u>David W. Richards (Arecibo)</u>.

<u>Thomas Gold (Cornell University)</u>: heard the Crab pulsar was slowing down, and calculated how much energy it would give off if it were a neutron star. This matched the amount of energy which had been calculated that the Crab Nebula needed to be supported, and so confirmed the idea that the pulsar was a neutron star.

<u>Jeremiah P. Ostriker (Princeton University)</u>: explained his model of a rotating magnetic field to supply the energy to the nebula.

<u>Donald J. Taylor</u>, W. John Cooke, and M. Disney (at Kitt Peak): discovered the optical pulsar.

<u>Paul Horowitz (Harvard University)</u>: showed the optical pulse and interpulse.

<u>E. Joseph Wampler (Lick)</u>: showed his stroboscopic identification of the pulsing star.

<u>E. Gross (Princeton)</u>: discovered the glitch. The existence of glitches was confirmed two years later at Lick Observatory. A glitch is explained by Ruderman as a starquake.

Ultraviolet, infrared, x-rays, and gamma rays studied from above the ground. X-ray rocket, ultraviolet Orbiting Astronomical Observatory shown. X-ray telescope to make images shown. Infrared balloon shown at launch at Palestine, Texas. Launch takes place at dawn, when the air can be especially still.

The Crab pulses even more strongly in x-rays and gamma rays.

Scientists search for gamma rays by trying to observe electrons that are travelling faster than the speed of light in air. (The electrons, of course, are not travelling faster than the absolute limit of the speed of light in a vacuum, which we call c.) This kind of radiation is called Cerenkov radiation. The Crab cannot be detected with the current level of technology in this part of the spectrum.

The Crab gives off 25,000 times the power of the sun.

<u>Stirling A. Colgate</u> explained what went on in the first seconds of a supernova explosion. He has set up an automatic supernova search in New Mexico, hoping to find supernovae only a few hours after they go off. The telescope can slew rapidly around the sky. (<u>Slew</u> means to swing around, and in astronomy is used to imply a rapid motion as opposed to a slower motion used to make fine adjustments in the pointing of the telescope.) By observing a supernova in its first moments, he hopes to be able to study the formation of heavier elements, and decide between competing models for what goes on in a supernova. [Note: The equipment continued to be developed but then funding ran out and the project was never brought to fruition. JMP]

-71-

TRANSCRIPTS AND NOTES ON MOVIES

The following are notes from the movie, <u>The Universe</u>: <u>Man's Changing Perceptions</u> (Preston Cinema Systems).   A transcript follows.

History of astronomy. Egyptian cosmology. 6th century cosmology of the monk named Cosmas. Points out that spherical earth was well established by the time of Columbus, but that flat-earth maps still existed. Just as Columbus never reached the edge of earth, neither could we reach an edge of the universe.

Olbers' Paradox: presents both solutions: expansion and age.

Kitt Peak: shows general view and 4-meter telescope in particular.

Interstellar matter: shows Orion, Rosette, Dumbbell, Lagoon (and M16 and Trifid, whose names weren't mentioned). Stresses that they are minute compared to the scale of the universe. Gives scale of universe in terms of travel time of light.

Shows map of Milky Way, Andromeda, Sculptor, elliptical in Cassiopeia, Triangulum with blue giants in its arms. Defines Local Group, shows clusters in Leo and Hercules.

Return closer to home to study the sun. Solar telescope at Kitt Peak. Time lapse H-alpha flare photographs (including photographs from Big Bear). Spectrum, including Fraunhofer lines.

Objective prism spectrum of Orion.

Defines photons, Doppler shift.

Shows Hubble and defines Hubble's Law. (Spectra are very nicely shown with suitable colors.)

Expanding universe. 15 billions years ago, just hot gas. Fireball, which gave way to our relatively cold and empty universe.

Steady state.

Shows Owens Valley radio telescopes.

1054 supernova and suitable Chinese drawings, Crab Nebula.

1968: pulsar discovered there.

Shows M82 in color and red differentiated (unfortunately mentions explosion, which is no longer thought to be the case).

3° black body: 1965, Penzias and Wilson and their telescope.

Big Bang confirmed.

"All that remains is a tepid sea of noisy radio static. So we are evolving from a fireball into an unknown future."

TRANSCRIPTS AND NOTES ON MOVIES

The following is a transcript of the movie, The Universe: Man's Changing Perceptions (Preston Cinema Systems).

> "The life of a star begins in the lake.
> It goes forth from the water;
> it flies upwards out of the previous form.
> It is the life of the stars."

Three thousand years have passed since these words were written down in the Nile Valley of Egypt. But they still convey to us a feeling that men in those times, closely hemmed in by the sun and stars, saw the heavens as an extension of their everyday experience. As the bounds of the known universe have grown, so has our own importance diminished. This film is about these changing perceptions of the cosmos.

In the times of the pharaohs, Egyptian cosmology was concerned more with worshipping the heavens than inquiring into their nature. The motion of the sun and stars across the sky was perceived as a sign of life. Here, in the Egyptian desert, we find the ancient records of these perceptions. Inside the tomb of Ramses IV, artists represented the heavens as a family of gods and goddesses. The god of the air, Shu, held up the sky, symbolized by his daughter, Nout. Her arched body separated the sky from the surrounding primeval waters. Every evening, Nout swallowed the sun and the world would be left in darkness. During the night the sun god would travel within the body of the goddess. In the morning he would be reborn, casting the rosy colors of birth into the canyons of the Nile.

By the sixth century Christianity had replaced paganism in Europe, and celestial objects were no longer worshipped as gods. Still, this century would see one of man's most absurd cosmologies, invented by an Egyptian merchant, turned zealous monk, named Cosmas. Inspired by the scriptures, Cosmas invented a rectangular picture of the world. Reminding us of his seriousness, Cosmas lectured: "Cease O ye wiseacres! prating worthless nonsense and learn at last, though late, to follow the divine oracle and not your own baseless fancies." The world of Cosmas illustrates just how far astray an unbridled imagination can wander. Cosmas explained the rising and setting of the sun orbiting around a huge mountain in the far north. Cosmas offered disbelievers numerous proofs. To those who still believed in the flaw: if rain falls downwards on us, then it must fall upwards on the earth's opposite side. Of course, this is contrary to nature, said Cosmas, and so the earth cannot be round.

Although the spherical earth was well-established by the time of Columbus, flat-earth maps continued to be popular. Such misconceptions undoubtedly contributed to the anxiety felt by the men who sailed westward with Columbus from Spain in 1492. Columbus did not set out to discover the New World. Promising gold and spices for the Queen, and a million converts for the Spanish Church, he sailed for Asia. Setting foot on the New World, he rediscovered for Europe what Asian tribes had come upon centuries before. To his death, Columbus insisted he had found China; that Cuba was Japan and Panama was Malaya.

Despite his magnificent error, Columbus began a great process of exploration--first covering the earth and then continuing into the depths of space. And we can draw a parallel to this voyage in cosmology. Just as Columbus never reached an edge to the earth, a space traveller would never reach an edge to the universe. Unlike the world of Cosmas, the universe has neither a center nor an edge.

Scientific cosmology began with men who wondered, not about the bright stars, but about the dark spaces between. These men asked, "Why is the sky dark at night?" The astronomers who first posed this question generally accepted the 16th-century belief that the universe is both infinitely large and eternally unchanging, filled with a limitless number of glowing stars.

The 19th-century astronomer, Heinrich Olbers, realized that this cosmology was incompatible with a dark night sky. For anywhere we look in the sky we should see a glowing star, and therefore the sky should be burning with light, as bright and hot as the sun. What Olbers discovered was that the dark night sky contradicted the idea of a static, infinite universe. The resolution to Olbers' paradox came with the discovery that the universe is not eternal and immutable, but goes through a process of birth and decay, not unlike our own mortality. And it is the age of the universe and the mortality of the stars that limits the brightness of the night sky.

This is for two reasons: first, that the stars simply don't have enough energy to fill the universe with light--they burn themselves out long before the universe becomes even lukewarm. The second reason is that the beginning of the universe represents a horizon beyond which we cannot see--because the light from more distant regions has not yet reached us.

Like giant eyes fixed at the sky, these are the instruments which have extended our perceptions to the heavens. This series of telescope domes crowns the top of Kitt Peak, in southern Arizona. The largest of these light-gathering instruments has a mirror 4 meters in diameter; it gathers 200,000 times the light of the unaided eye. Its massive tracking system compensates for the earth's rotation and keeps the telescope pointed at its target for the many hours necessary to record a long photographic exposure.

These great collectors of light have opened the heavens, revealing to us their nature. The evidence we see is of a living universe which evolves and changes, and not the static universe of Olbers or the rigid box of Cosmas. Witness to change is given by the vast clouds of glowing gas called nebulae. The Orion Nebula is a vast cloud of hydrogen gas in the process of slowly condensing into stars. Hot stars born into this cosmic nursery are responsible for its brilliance. The Rosette Nebula came into being as a star threw off part of its atmosphere in a great explosion. The Dumbbell Nebula, like the Rosette Nebula, also is the discarded mantle of a star. The Lagoon Nebula is split into two great clouds by a dark gulf of dust extending trillions of miles.

Enormous as these nebulae are, they are absolutely minute relative to the universe as a whole. It's extremely difficult for us to relate to the scale of the universe since the distances and times we're used to relate to activities on our own scale. Perhaps because we are familiar with long periods of time--at least in the context of history--it's convenient to specify astronomical distance in terms of the travel time of light. Since this film began, light leaving the sun has just passed through the orbit of Mars. But it will be 4 years before that light reaches the nearest star. And so the distance light travels in a year's time, called the light year, has become a common, although somewhat small, measure in astronomy. The light year is 6 trillion miles.

This glowing cloud of a hundred billion stars for us is home. Our sun is one of the multitude of stars which form a glowing island in space called the Milky Way Galaxy. Our galaxy is so vast that light takes 100,000 years to cross from one end to the other. The light we are now receiving from the most remote stars in our galaxy began its journey when early Neanderthal man still roamed the earth. Floating in this sea of stars, it's extremely difficult for us to perceive the overall structure of our own galaxy. The easiest way to visualize this structure is to look at the nearest galaxy similar to our own, the galaxy in Andromeda.

Very nearly a twin to our Milky Way, Andromeda has a bright round nucleus, surrounded by a fainter disc of stars. Located about 20 diameters of our own galaxy away, the light which we now see from Andromeda began its journey 2 million years ago. Besides Andromeda, the Milky Way has almost two dozen close neighbors. Less than a million light years away is this spiral galaxy in the constellation, Sculptor. Nearly two million light years away lies this small elliptical galaxy in the constellation Cassiopeia, which can be resolved into a swarm of individual stars. And in the neighborhood of Andromeda, just over 2 million light years away, is this galaxy in Triangulum--whose twisting arms are strewn with blue giant stars.

Galaxies like these tend to cluster in groups; we and our neighboring galaxies form what is called the Local Group.

This small group consists of 4 galaxies in the constellation Leo. In contrast, this enormous cluster of galaxies in Hercules contains over 10,000 members. Nearly every extended point of light we see here is a galaxy and each galaxy on the average contains over a billion stars. It is one of the largest organizations of matter in the universe.

The faint light coming from distant galaxies carries information about the structure of space itself. To see how astronomers gather this information, let us return closer to home, to study our own star, the sun. Looking like a work of modern sculpture, the solar telescope at Kitt Peak is the world's largest instrument for studying the sun. The mirror atop the vertical tower tracks the sun as it moves across the sky and reflects the sunlight down the slanted shaft, 500 feet.

At the bottom other mirrors focus the light and direct it into laboratories where is it analyzed.

These time-lapse photographs of the sun have been taken in the red light emitted by hot hydrogen gas. Violent flares reveal the sun's turbulent, stormy atmosphere. Iron, hydrogen, calcium--these and the other elements found on earth are swept about the sun's surface in the form of hot gases. In the sun's atmosphere, these elements absorb certain colors from the light streaming out of the interior. These absorbed colors are the key to understanding not only our sun, but also the nature of distant stars.

In order to make this process of absorption visible, astronomers break down the sunlight into its component colors, or spectrum.

Here, cutting across the rainbow of colors are the dark absorption lines. These lines are the signature of elements, 93 million miles away, in our sun. The same processes are at work in more distant stars. The light from each of these stars in the Orion Nebula has been decomposed into its spectrum. Although the absorption lines of a distant star are caused by the same elements found on earth, their position in the spectrum can change. This is a result of both the star's motion and the nature of light.

Light is made up of small bundles of energy called photons. Our eyes see the energy of photons in terms of color. If a star moves toward us the photons we see have more energy and the star appears more blue than it if were stationary. Likewise, a star receding from us appears more red since the photons reaching us have less energy. These color shifts are the key to measuring the motion of distant stars. Astronomers simply compare a star's absorption lines with their standard set. The lines of an approaching star are shifted toward the blue and the lines of a receding star are shifted toward the red.

In the 1920's, the astronomer Edwin Hubble used the shifting of absorption lines to relate the motion of remote galaxies to their distance. After studying the spectra from many faint galaxies, Hubble announced his great discovery: That distant galaxies are all moving away from us, and the speed at which they recede is proportional to their distance.

This relatively nearby galaxy in the constellation Virgo is 78 million light years distant, and is moving away from us at 1200 kilometers per second; while the faint galaxy in the constellation Hydra is almost 50 times more distant, and is therefore moving away 50 times faster; about 60,000 kilometers per second.

In order to understand this rushing away of distant galaxies, we have to abandon the view that we are somehow fixed in the center of the universe, with all other galaxies moving away from us. The universe itself is expanding, and as the fabric of space-time stretches out, the galaxies all move away from one another. An observer in some far off galaxy would see the Milky Way as well as all of his other remote neighbors receding from him. This is Cosmic Democracy.
These glowing areas are created in the physics laboratory. If we

could look far enough back into time, this is how the universe would appear, fifteen billion years ago. No stars, planets, or galaxies; just a featureless broth of elementary particles and radiation. Then the universe began to fly apart; the dense, hot fireball gradually gave way to our relatively cold and empty universe.

But did the Big Bang really occur? An opposing cosmology, called the Steady State theory, says that matter is continually being created in space, but in such small quantities that the process escapes detection. This newly created matter fills in the voids left by the expanding universe, so that the appearance of the universe always remains the same.

The crucial information which would allow cosmologists to choose between these two competing theories came from an observation as fundamental as the night sky being dark. Unlike the evidence provided by the dark sky, this evidence, carried by radio waves, lay beyond the vision of our eyes. Streaming in from space, radio waves are being collected with these giant antennas in the Owens Valley of California. The waves we now receive began their journeys at diverse times; some originating long before the sun and earth were formed, bring with them a glimpse of how the universe began. Other signals echo more recent cosmic history.

In 1054 the first signals arrived from an event which had occurred 5,000 years earlier. It was a supernova; a star exploding like an enormous nuclear bomb. The event was recorded by Chinese astronomers as a brilliant, but temporary, addition to the heavens. This supernova was the parent of the Crab Nebula, a great cloud of gas still expanding at 1,000 kilometers per second. In 1968, radio astronomers discovered that the center of the Crab Nebula contained an extraordinary object, emitting pulses of energy at precisely regular intervals. This byproduct of the supernova, called a pulsar, is thought to be the collapsed core of the exploded star, spinning at 30 revolutions per second. The compressed material of the pulsar is so dense that a teaspoon of its matter would weigh a hundred million tons on earth.

Many strong radio sources are located far outside our own galaxy. This galaxy, called M82, appears in the midst of a titanic explosion. In the red light emitted by hot hydrogen gas, we see that an enormous jet of hydrogen, more than 10,000 light years long is pouring from its nucleus. This violently expelled matter, curving along the galaxy's magnetic field, is the source of intense radio noise.

We are now turning our antenna towards a galaxy whose place in cosmic evolution is a mystery. Called Cygnus A, it is one of the strongest radio sources in the sky. We receive more radio power from this inconspicuous pair of clouds than we do on the average from our own sun, even though the waves from Cygnus A have taken 600 million years to reach us compared to 8 minutes for sunlight. It is still not known from what process Cygnus A draws its energy, or what role these violent eruptions play in the evolution of a galaxy.

In 1965 these two scientists, Arno Penzias and Robert Wilson, discovered evidence of the greatest violence of all, the Big Bang. They found that the earth was immersed in a sea of radio emissions, called microwave radiation. These surviving relics of the early universe come uniformly from all regions of the sky. They confirm the Big Bang hypothesis: that about 15 billion years ago the universe was in the form of an enormous fireball filled with intense radiation. As the universe expanded, the hot brilliant light was redshifted, losing energy, until now all that remains is a tepid sea of noisy radio static.

And so the microwave background seems to confirm that the universe is evolving out of the ashes of an immense fireball, into an as yet unknown future. This revelation of mutability cannot reflect badly on the authors of the ancient papyrus which stated:

"The life of a star begins in the lake.
 IT GOES FORTH FROM THE WATER,
 IT FLIES UPWARDS OUT OF THE PREVIOUS FORM.
 It is the life of the stars."

Taken as a metaphor, the life of a star does begin in a lake--the sea of radiation which accompanied the birth of the universe and surrounds us still.

****

TRANSCRIPTS AND NOTES ON MOVIES

The following is a transcript of the movie, The Creation of the Universe, by Timothy Ferris, winner of the 1985 AAAS-Westinghouse Science Journalism Award.

PART I: THE MACROWORLD

Einstein dreamt of finding a unified field theory... a single equation that might account for every fundamental process in nature, from the jostling of atoms to the wheeling of the galaxies.

Today, science is close to fulfilling Einstein's dream. The nucleus of the atom is yielding up evidence of an elegant simplicity underlying the wild diversity of the universe. New unified theories are being written that reveal traces of this primordial simplicity. The 1984 Nobel Prize in physics was awarded for experiments that confirmed the first of these new theories.

Experiment and theory alike indicate that the universe began in a state of perfect simplicity, evidence of which was burned into the heart of every atom in the heart of the big bang at the beginning of time.

MICHAEL TURNER: We have a pretty good understanding of the history of the universe from a hundredth of a second after the big bang until today, 15 billion years later. It's pretty remarkable that I can say this, and it's even more remarkable that I can say it and the men in white coats don't come and pull me off the stage. Part of the history we have pretty much nailed down, because there are fossils and relics that are left behind that tell us that our theory is right.

MURRAY GELL-MANN: Cosmology, it turns out, provides, in a way, a sort of testing ground for some of the ideas of elementary particle physics. We can't observe the early universe, but we can observe its consequences in the universe of today.

ALLAN SANDAGE: What's it like out there? I don't know what it's like out there. It's cold. It's impersonal. It--it is the machine, if you like to put it that way, that has created you. Now, by that, I mean the following: every single atom in your body was once inside a star. We are all brothers in that sense.

To know the atom, it seems, we must know the universe, and to know the universe we must know the atom. In a confluence of large and small that would have gladdened Einstein's heart, the search for simplicity is bringing science face to face with the ancient enigma of creation.

When I was a boy I used to bring a telescope down to the shore, point it up into the sky and look out into what I took to be the universe. I had read in the astronomy books that what we see in the sky is the past. The Big Dipper, for instance, is a star cluster 75 light years away. That means it takes light from the stars of the Big Dipper 75 years to speed through space and reach our solar system. If you're lucky enough to get to be 75 years old, you can see the Big Dipper just as it looked on the day you were born. This star cluster is further out. It's in the constellation Hercules and it's 27,000 light years away. That means that we see it as it looked 27,000 years ago, in the days when there wasn't a human city anywhere on the face of the earth, and the bow and arrow hadn't been invented.

One way to study the history of the universe is to look out into deep space. But the universe, after all, isn't all out among the stars and galaxies. It's also right here at home. We live in the universe. And ordinary objects are full of clues to cosmic history. Every stone on the beach is a galaxy of atoms and those atoms have a history that goes back longer than the history of the earth. Before this earth was formed, those atoms were adrift in interstellar space. Before that, some of them were incorporated into ancient stars. The subatomic particles that make up these atoms can trace their lineage back 15 billion years, to the beginning of time.

In this program, we'll explore the history and the origin of the universe. We'll look down into the minuscule world of the subatomic particles and out toward the cosmological frontiers of space and time. Our hope is to learn something about how everything got to be the way it is. Clues to that riddle are all around us.

Skyscrapers cluster in the Wall Street district of downtown Manhattan. For more than a mile to the north, there are no tall buildings... and then, in midtown Manhattan, the skyscrapers sprout up again. It's a natural development, its destiny laid down hundreds of millions of years ago, when molten lava flowed through the Hudson River valley. The bedrock on which the city is built was crumpled and wrinkled by geological processes, and the result was a pair of underground mountains. Since skyscrapers must be rooted in bedrock, the swayback form of the Manhattan skyline traces out the contours of a subterranean mountain range. It's here that we begin our exploration of the realm of the atom and the depths of the past.

Times Square is about as unnatural a piece of real estate as you can imagine. And yet it, too, has been shaped to a considerable extent by its natural history. The planetary processes that shaped it are by and large to imperceptibly slow for us to notice on the human time scale. But imagine that we could walk into the past at an accelerated rate. Then we could see how the geology of Times Square was shaped.

Let's try it. Let's walk into the past at the rate of a century per step. A century ago Times Square, known then as Longacre Square, was a dark, dangerous place, a haven for muggers. Its transformation into the Great White Way--a brighter if not substantially safer place--was due to the applied physics of Thomas Edison and his invention, the electric light.

Two hundred years before that, Times Square was farmland. The Hopper family raised cabbages at Broadway and 50th Street. A few more steps into the past and Broadway was an Indian trail. Only seventy steps--7,000 years ago--and Manhattan had yet to be discovered by the Indians. A little over a hundred steps, and we've come to the epoch of the mastodons. Mastodons once grazed here, in Central Park.

Only a 150 paces suffices to take us back to the most recent Ice
Age, when Manhattan was buried under a sheath of ice over a thousand
feet thick.   This boulder is what the geologists call a glacial
erratic.   It was pushed down here from upstate by the advancing
ice--and you can still see how these exposed rocks were polished by
the glacier as it moved south.   But the rocks themselves are millions
of years old.   To walk that far back in time at a century a step,
you'd have to walk to the North Pole.   And the earth is billions of
years old.   To go that far, you'd have to walk clear around the world.

Scientists deal with big numbers like these by using what they
call exponentials.   For instance, powers of 10, 10 to the second is 10
times 10 or 100.   10 to the 3rd is 10 times that or 1,000.   This time,
let's imagine that we can walk deeply into the past exponentially, by
letting each step count for 10 times as much as the step that preceded
it.

A month ago, a year ago, ten years ago, one hundred years
ago, ten to the third, a thousand years ago--and the city vanishes.
Ten thousand years ago Manhattan was still awash with the melt of the
Ice Ages.   For a million years before that, the glaciers advanced and
retreated, cutting the great furrows that filled with the run-off
water.   One of them became the Hudson River.   One hundred million
years ago, Manhattan lay at the floor of the Cretaceous Sea.   Prior to
that, North America was still connected to Africa.   Volcanoes spewed
forth lava in which the skyscrapers of Manhattan were one day to take
root.   One billion years ago, New York City was the site of a mountain
range as imposing as the Alps.

Four to five billion years ago, the earth was still condensing
from fragments of rock circling the sun.   Before that there was no
earth, just a drifting interstellar cloud from which the sun and its
planets were to form.   Tennyson wrote of the abyss of time:

> "...O earth, what changes hath thou seen!
>   There where the long street roars hath been
>   The stillness of the central sea.
> The hills are shadows, and they flow
>   from form to form, and nothing stands;
>   They melt like mist, the solid lands,
> Like clouds they shape themselves and go."

The earth is old, but older still are the atoms that compose the
earth, and the air, and you and me and this tree, and we can
investigate the depths of cosmic history by looking at nature on the
atomic and subatomic scale.

The more closely we scrutinize the universe of the very small,
the more evidence we find of the past.   The pollen blossoms caught in
this tree are young but their color and form were generated by genes
that are older than the plant itself.   The genes, in turn, are built
upon DNA molecules.   The molecules become visible at a magnification
of ten million times.   Each is a library of genetic information
accumulated over eons in the evolution of life on earth.

Moving in more closely still, we can see the carbon atoms of which the DNA molecules are made. They're even older than the earth. The outer particles of the atom are patrolled by a shell of negatively-charged particles--the electrons. Photons carry electromagnetic force between the electron shells. We pass through the inner shell of the carbon atom.

We are approaching one of the oldest and most magnificent structures in nature--the nucleus of the atom. The nucleus is made of protons--here colored orange to indicate their positive electrical charge--and electrically neutral particles, the neutrons. These nuclear particles are, in turn, made of trios of even older and more fundamental particles, the quarks. We've reached the realm of the nuclear forces. The weak force mediates the process of radioactive decay while the strong force binds the quarks together, weaving webs of energy into the forms that we call matter.

Most of what we know about subatomic particles has been learned by accelerating the particles to a high velocity, colliding them with one another and studying the debris that comes flying out. One physicist compared it to smashing a pair of elegant Swiss watches together and then trying to figure out how they were designed by looking at the cog wheels and screws that would come flying away. It's not the world's most elegant or subtle method, but it works.

This was the world's first particle accelerator. It employed principles that are still in use today. Positively charged nuclear particles, protons, were injected into this little ring, where they were subjected to a powerful electromagnetic field. Since protons are positively charged, they were pulled around toward the negative side of the field. Then, the polarity of the magnetic field was reversed, hurrying the protons around here. By repeating the process, you could get them going at a pretty good clip, until they spiralled on out to the edge of the ring and finally collided with this target here.

This prototype was built in Berkeley, California in 1930. It won its inventors a grant of $500.00. They used the money to build a second accelerator twice its size. That one was followed by another, still larger accelerator. Particle accelerators have been getting bigger and bigger ever since.

The large the accelerator, the smaller the scale to which it can probe the fundamental structures of nature. And I'm standing in the midst of one of the world's largest particle accelerators--it's Fermilab, on the Illinois plains. It works much like the first accelerator by using electromagnetic force to whirl particles around the ring, but the ring has gotten larger. At Fermilab it's three miles in circumference.

In a tunnel smaller than a fire hose, buried beneath the earth, particles are accelerated to nearly the speed of light and then collided with a fixed target. The resulting explosions, recorded in detector tracings like this one, reveal a wide variety of subatomic particles that interact by means of four fundamental forces--gravitation, electromagnetism, and the weak and strong nuclear forces. The way physics sees it, particles and forces are the authors of events in our world, from the exotic to the everyday.

"Hi there, sports fans! This is Biff Burns."

"And this is Steve Boscoe and it's a beautiful day for baseball. Thanks to the weak nuclear force, the sun is busy releasing energy from the nuclei of atoms making it a balmy seventy degrees here in Dodger stadium."

"The pitcher is on the mound, and we're all set to get underway."

"He really got a piece of that one."

"He sure did, Biff. But the earth's gravitational force warped space so much it looks like an easy out."

"Yup, he caught it all right."

"Well, it's not so much that he caught it. It's more that the electromagnetic field set up by the atoms in his glove surrounded the electromagnetic field of the ball."

"Well, whatever you say."

"Out. Well let's see--who's up next? Looks like Reggie. The pitcher is looking in to get his signal."

"Wow! What a wallop."

"He just about knocked the quarks out of the ball. It's out of here! A home run!"

"And this is Biff Burns saying until next time this is Biff Burns saying so long and this is Steve Boscoe rounding third and being thrown out at home."

The baseball and the bat are mostly empty space. Their solidity is an illusion created by the electromagnetic force field that binds their atoms together. Viewed on the subatomic level, a home run begins with the exchange of photons between atoms in the bat and the atoms in the ball. The ball is repelled and flies away. We credit the home run to the batter, but the fundamental force responsible is electromagnetism. Infinite in range, it's electromagnetism that brings us light from the sun and stars.

The weak nuclear force helps power the sun and presides over the phenomenon of nuclear decay. Tremendous amounts of energy are bound up inside the nucleus of each atom. Some nuclei are unstable and can't contain their energy forever. When they decay, it's the weak force carried by particles called weak bosons that governs the process.

The strong nuclear force binds quarks together to make protons and neutrons. Without it, there would be no atoms, and the universe would be a quark fog. The strong force is carried by articles that the physicists call gluons, because they act like the most perfect imaginable glue.

Gravitation, the universal attraction of all massive particles toward one another, is the weakest of the four forces. But gravity has infinite range, and it always attracts, never repels. This single-minded dedication makes gravitation able to hold the planets, the stars, and galaxies together.

But why are there four forces, and why do they differ so profoundly in character? Einstein tried in vain to find an answer to that question. He sought what he called "a simplified and lucid image of the world"--a single principle that would account for the baffling differences between the forces, and the great variety of particles. "What really interests me," he said, "is whether God had any choice in the creation of the universe." He failed in that effort, but he never lost his faith that as he put it--"God is subtle, but not malicious"-- that nature, though difficult to understand, ought at the root to be simple and beautiful. Albert Einstein spent the last years of his life in this office at the Institute for Advanced Study in Princeton, New Jersey working on the search for a unified theory of the forces of nature. He never found it. The equations remained unfinished on the blackboard on the day he died.

Today, the equations look a lot different, and they go by fancy names that Einstein never heard of, like "supergravity" and "supersymmetry" and "superstring" theory. But their goal is the same, to draw the disparate elements of physics together into one whole.

Physics today is a patchwork. There are theories that account for each of the fundamental forces, and they do so marvelously well. They can predict the behavior of an electron spiraling away from an exploding galaxy or the behavior of these photons arriving here from the sun. But they have little to say about why the forces of nature differ so curiously in character. Why, for instance, is there a positive and negative electrical charge, but no such thing, so far as we know, as negative gravity? Why do the strong and weak forces function only within the nucleus of the atom, while electromagnetism and gravitation are infinite in range?

The search for simplicity--the hope of identifying the fundamental force or a basic building block of matter, runs all through the history of physics, since the days of the ancient Greeks. Atomic theory was already an old idea when Plato debated its merits with his pupil Aristotle. Aristotle proposed that there were two basic forces--he called them levity, the tendency of light substances to rise, and gravity, the tendency of heavy objects to fall. The Roman poet Lucretius popularized atomic theory:
    "Beautiful is the world created by the atoms. A wedding ring wears thin with the passage of the years, yet we never see flecks of gold departing from the ring, for the gold is made of tiny atoms...."

TRANSCRIPTS AND NOTES ON MOVIES

"Blasphemy! Atheism!" Amid the spiritualism of the dark ages, atomic theory, regarded as materialistic, was forgotten, only to be revived in the age of Isaac Newton, who saw the world as composed of atoms marshalled by the force of gravity. "The universe is like ... a perfect machine. It is the force of gravity which holds the moon and planets in their orbits."

In the centuries that followed, experiments by Robert Boyle, John Dalton and others revealed that, as Ernest Rutherford put it, "the atom is a complex aggregate, not a simple entity...." Rutherford found positively charged particles--protons--in the nucleus of the atom, and James Chadwick found neutrally charged neutrons there.

James Clerk Maxwell, in the first modern unified theory, showed that electricity is but an aspect of electromagnetic force. For a time it seemed that there might be just two fundamental forces, electromagnetism and gravity. But then the force picture became more complex when Hideki Yukawa and Enrico Fermi and others identified two previously undetected forces acting within the nucleus--the strong and weak nuclear forces.

With the twentieth century came the identification not only of new forces but of scores more particles. Einstein proved that energy itself is composed of particles--the quanta. Wolfgang Pauli found an odd particle called a neutrino, and the existence of the even stranger antimatter was established by Paul Dirac. The list kept growing. There were mesons and muons, and pions. Said Fermi, "If I could remember the names of all these particles, I would have been a botanist."

Hopes of finding an ultimate building block of matter were raised anew when Murray Gell-Mann proposed that protons and neutrons are made of still smaller particles that he called "quarks." The concept of force was simplified as well, when Sheldon Glashow, Abdus Salam and Steven Weinberg showed that electromagnetism and the weak force are aspects of a single, electroweak force. Experiments by Simon van der Meer and Carlo Rubbia confirmed the electroweak theory.

More ambitious "grand unified" and "superunified" theories followed, and by the mid 1980s, hopes ran high that physics might be within reach of an ultimate unified theory, a single equation able to explain the toilings of quarks and stars.

LEON LEDERMAN: The trouble we're in now is that the standard model, the standard picture, is very elegant, it's very powerful, it explains so much... but it's not complete. It's incomplete. It has some flaws. And one of its greatest flaws is one which is hard to explain. It's an aesthetic flaw. It's too complicated. It has too many arbitrary parameters. We don't really see the creator twiddling 20 knobs to set 20 parameters to create the universe as we know it. That's too many. Ever since the Greeks started us on this road to understanding the atoms, the fundamental building blocks of the universe, we've had this prejudice that there's something simple underneath all of this. And six quarks and six leptons and their antiparticles, and their coming in different colors and in different charges, is too complicated. And there's a deep feeling that the picture is not beautiful. And that drive for beauty and simplicity and symmetry has been an unfailing guidepost to how to go in physics.

STEVEN WEINBERG: We haven't come to the bottom level yet. But as we approach it, we pick up intimations of an underlying beautiful theory whose beauty we can only dimly see at the present time. We don't know. We don't know that it's true. We don't know that there really is a beautiful underlying theory. We don't know that as a species we're smart enough to learn what it is. But we do know that if we don't assume that there is a beautiful underlying theory, and assume that we're smart enough to learn what it is, we never will.

JOHN WHEELER: To my mind, there must be at the bottom of it all an utterly--not equation, not an utterly simple equation--but an utterly simple idea. And to me, that idea, when we finally discover it, will be so compelling, so inevitable, so beautiful, that we will all say to each other, "Oh, how could it have been otherwise."

The unified theories suggest that nature would function more simply under conditions of extremely high energy. Take electromagnetism and the weak force. At normal energy levels, they seem very different. Electromagnetism is conveyed by photons. Photons are lightweight and they can travel for vast distances. But the weak force is a different matter. It's carried by "weak bosons." They're heavy, and can travel only very short distances before they exhaust themselves and decay. That's why the weak force is limited in range to the nucleus of the atom.

But the theories say that the situation would change if we could turn up the heat. Fueled by the ambient energy, a new particle called the"Z " would appear and would be capable of knitting together electromagnetism and the weak force. In this computer simulation, we'll watch as a Z particle decays and recombines to forma photon, carrier of electromagnetism. The photon, in turn, decays to form a pair of weak bosons, carriers of the weak force, and the bosons transform themselves back into a Z. What had been two forces is now one **electroweak** force.

One way to test the theory was to look for Z particles. Like salamanders, the mythological creatures that dwell only in fire, Z particles thrive only under conditions of intensely high energy. The universe today is too cold for Z particles to exist for long: they would find it chilly even in the interior of a super giant star.

When the electroweak theory first predicted that there ought to be such a thing as a Z particle, no laboratory on earth could summon up enough heat to test that prediction. It wasn't until 1983 that science managed to fire up a spark hot enough to summon up the Z particle if it existed. It happened in a laboratory here on the borderline between Switzerland and France.

The site was CERN, an international laboratory administered jointly by thirteen European nations. Like other giant particle accelerators, CERN consumes as much electricity as a small city, but it manufactures nothing. The six thousand people who work here are engaged in pure research.

To achieve the high energies at which Z particles would appear and electromagnetism and the weak force would begin to merge, scientists and CERN tried an unprecedented experiment. They sought to collie matter with antimatter.

It all starts here. The particles that are accelerated are protons. They're easy to come by. There's at least one proton in the nucleus of every atom in the universe, and they're economical. This one bottle of hydrogen gas contains a year's supply of protons for the CERN accelerator. The gas is emitted in infinitesimal little puffs through these computer-controlled valves and emerges into this pipe, the headwaters of the entire accelerator. Those tiny puffs of gas each contain as many protons as there are stars in the Milky Way galaxy.

This steadfast old generator cranks out nearly a million volts. The power is used to set up an electromagnetic field in this chamber. In the field, the negatively charged electrons orbiting the nucleus of each hydrogen atom are stripped away, leaving the denuded, positively charged proton. The electrons remain behind and the protons speed off toward the main accelerator.

For years, scientists had been accelerating protons in the giant CERN ring and colliding them with stationary targets. But now, for the first time, anti-protons--rare particles identical to protons, but with opposite charge--were manufactured and stored in this antimatter collector. Then the anti-protons were injected into the main ring, hurtling in the opposite direction as the protons.

When matter encounters antimatter, the result is mutual annihilation. The impact between a single proton and anti-proton releases, for a tiny fraction of a second, more energy than the electrical output of all the power plants in the world. Subatomic particles come reeling out of that tiny explosion and are captured in the onion skin layers of this giant particle detector.

On the 28th of May, 1983, traces of a Z particle were detected in the debris emitted by a proton-antiproton collision. The electroweak unified theory had been confirmed. Carlo Rubbia of CERN shared the 1984 Nobel Prize in physics for having conceived of the idea of colliding matter with antimatter.

RUBBIA: What we are really doing here--we are making little bangs. We are concentrating, in a very small volume of space for a very short period of time, enough energy, density, so that we can revive, or replace so to speak, on a very modest scale, what was really the state of affairs of the universe as a whole. The major progress in science has been made, I believe, by Galileo Galilei when he brought experimental science to its right level of importance. And I still believe that to a greater extent, our scientific progress is made through experimental science. It's not because I am an experimental scientist. Just because I believe that there is always a final verdict--the final guidance--which comes from the physical phenomenon.

With the electroweak theory a success, scientists went on to compose the so-called "grand-unified theories." These theories say that, at sill higher energy levels, three of the four forces might function as a single force. And here again, it's theorized that an exotic particle would do the work of unification. It's called the "X" particle.

As we watch, a gluon, carrier of the strong force, strikes an X particle and is transformed into a photon, the carrier of electromagnetism. Alternatively, a gluon striking an X can be transformed into a weak boson, a carrier of the weak force.

Particles of energy live on borrowed time; they gather themselves up from the stray energy in a vacuum--then ebb back into nonexistence. The grand unified theories make the startling prediction that not just energy, but matter may be temporary. They say that protons, the particles that form the heart of every atom in the universe, are not permanent, as had been thought, but are destined to decay.

The grand unified theories paint in deepened colors the old lesson that birth implies death. It's the latest chapter in a scientific saga that began here in Venice, 375 years ago.

On August the 25th, 1609, Galileo led a procession of Venetian senators across the square and up to the top of that tower for their first look through his first telescope. Galileo was teaching just up the river at the University of Padua at the time. He was a respected teacher. Students flocked to his classes. He'd written a couple of good books. But his contract was about to expire, and he'd never made enough money to get by. He needed something to help his career. He found it. It was the telescope.

The senators were impressed. They granted Galileo tenure, gave him a promotion and commissioned telescopes for sighting telescopes at sea. But Galileo trained his telescopes on the sky. He saw that the moon has mountains as rugged as the Apennines, and that other planets have moons, that the Milky Way runs deep with pasturelands of stars, and these sights helped to convince him that the heavens are as substantial and changeable as the earth.

Galileo's observations provided enough evidence that stars and planets are worlds like ours, made of the same elements and functioning in accordance with the same physical laws. They drew together the realms of large and small, and married the earth to the universe.

The Venetians understood that nothing on earth lasts forever. They were citizens of a republic literally raised upon shifting sands, and supported by the risky business of ocean-going trade. But they thought that the stars, at least, were immutable.

Galileo's work changed that. By introducing into science the idea that the earth is part of the universe, he set the stage for the subsequent discovery that even stars live and die. Since Galileo's day, telescopes have looked out into space to scales Galileo never dreamed of, and particle accelerators have looked just as far into the world of the small. And everywhere, they have found change. If the grand unified theories are right, not even atoms last forever.

The city where Galileo glimpsed signs of the mortality of the heavens is itself in peril. Venice is sinking, its streets flooding with every winter's storm. The city cannot last forever, but it has company in that; neither, it seems, can the sun, nor the stars, nor the atoms they are made of.

Kamioka, Japan. Here, in a lead mine, an experiment is underway to learn whether protons are mortal. To minimize interference by high energy particles coming from space, the experiment is conducted nearly two miles underground. The unified theories predict that the average proton will last many billions of years, and that, by assembling an enormous quantity of protons, scientists can test the theories in the course of a year or two of careful observation. And that's what's being done here.

A pit has been excavated, rigged with sensitive detectors, and then filled with 660 tons of pure water. If a proton in any one of the billions of water molecules in the tank were to decay, it would send out a telltale flash of light. Computers monitoring the flash of light would bring the news to waiting scientists. If the results are positive, the experiment will confirm an unsettling hypothesis--that all the matter in the universe is a passing fancy, a stage in the ongoing dance of energy.

If electromagnetism and the weak force unite at high energies, and the strong force joins with them at still higher energies, then might not all the four forces ultimately be one? Some of the most inventive minds in physics are at work trying to solve that riddle.

Stephen Hawking ranks among the world's leading physicists, despite his having suffered, for more than twenty years, from a progressive disease of the central nervous system that has left him paralyzed and almost unable to speak. Hawking holds Isaac Newton's old chair as Lucasian Professor of Mathematics at Cambridge University. During a recent seminar at Caltech, his words were interpreted by a former student, supersymmetry physicist Nick Warner. Hawking's subject: the infancy of the universe.

WARNER (TRANSLATING HAWKING): "The boundary conditions are that the universe has no boundary." Sounds rather Zen, doesn't it? What Steven means by this is that spacetime is some compact, four-dimensional manifold. This is a general four manifold--some compact four-manifold M, with some metric on it, G Mu Nu, on the surface, and also some matter fields, Phi, on the surface.

MURRAY GELL-MANN: Do I understand then, that you want the mass of effective unification, of a Yang-Mills or super Yang-Mills theory, to be this low, like 10 to the 14th? Or can that M be something else?

WARNER (TRANSLATING HAWKING): It may be something else.

GELL-MANN: But in any case, the M is the mass characteristic of whatever physics is responsible for the transition?

WARNER (TRANSLATING FOR HAWKING): It is a mass. It generates--it generates the inflation cosmological constant. Steven just wants to add that you can also explain--you can also explain the arrow of time from this. He thinks that that would take much more time than we have. That's all.

No laboratory on earth can produce the energy at which the four forces would act as one. Only the fires of genesis were hot enough for that. The search for simplicity in the realm of the atom draws us out into the realm of the galaxies, and back toward the beginning of time.

PART II: MACROWORLD

We live in a major spiral galaxy that we call the Milky Way. It's home to the sun and a few hundred billion other stars. The galaxy belongs, in turn, to a cluster of galaxies; astronomers call it the Local Group, and the Local Group is part of the Virgo Supercluster, an archipelago of galaxies stretching across one hundred million light years of space.

The galaxies are marching away from one another as the universe expands. In a computer simulation based upon astronomical observations, we witness something never before seen by human eyes... the predicted scattering of the Virgo Supercluster, by the expansion of the universe, over the course of the next 50 billion years.

The expansion of the universe was predicted by Einstein's general theory of relativity, published in 1915. But the idea seemed so outlandish that Einstein himself rejected it. He introduced an extraneous term into the field equations to try to make his theoretical universe stand still. Later Einstein would call this modification of the theory, "The worst blunder of my career." Then in 1929, the American astronomer Edwin Hubble, without knowing of the relativity prediction, discovered that the universe is indeed expanding. Einstein and Hubble met in California in 1931, and celebrated Hubble's having found, at the telescope, what the mind of Einstein had conceived.

Astronomer Allan Sandage, once Hubble's pupil, has devoted much of his career to studying the expansion of the universe.

SANDAGE: It is not as if these galaxies are expanding into a space that's already there. The view is that space itself is expanding, carrying the galaxies with it. The expansion creates the space. The crucial analogy, first made by Eddington as long ago as 1930, just one year after Hubble had announced the expansion, was that you can conceptualize the thing as the two dimensional analog, by the surface of a balloon. You paint dots on the surface of a balloon and you blow it up. You put yourself on any dot. You seem to be the center and all the other dots move away from you. Now, take the air out of the balloon and look what the dots do. All the dots come toward every other dot and it you could take all of the air out of a perfect balloon the surface itself would go to zero; all the dots would be back at one place at one time. Every place is the center of the expansion. When you talk about this, the question that always comes, "well, can you find the center of the expansion?" Every place is the center of the expansion. There is no center to the beginning. Everything was back at--at one place and every place and every time was identical, in the beginning.

Palomar Observatory in Southern California. I'm sitting inside a time machine of sorts, an instrument capable of looking directly into the past. Down at the bottom of the tube far below me, there's a curved polished mirror that can gather as much starlight as all the eyes in a community of 20,000 people. That light is brought here to a single intense focus. Through that tiny window, one can peer out for billions of light years in space, and look back for billions of years into cosmic history.

You can harvest some pretty old light with a telescope as large as the 200-inch reflector at Palomar. This image just coming in now is of a galaxy 40 million light years away. That means we're seeing it the way it looked 40 million years ago--which was a long time, but it's only a fraction of one percent of the time that's elapsed since the beginning of the expansion of the universe.

If we were to plot our place in cosmic history, we might make a line representing time... starting with the big bang--the beginning of time, as best we can understand it--and stretching down some 15 billion years or so, to the present day. Here we are, in a galaxy today... and, we could have the vertical axis represent space.

Now we can only see events the light from which, travelling through space, has had time to reach us. And we can designate this by drawing what the scientists call a "light cone." The angle of the sides of the cone is defined by the velocity of light... the fastest way we know of that information can travel. A galaxy like this, at 40 million light years away is right here in our--in our own neighborhood. And pretty much all the other galaxies that we can see clearly lie quite close by on the cosmological scheme of things. If we look further out, the galaxies begin to get pretty dim.

Let's see if we can get a cluster of galaxies here. Each of those tiny little fuzzy dots--so small that you may find it difficult to see them--each of those dots is a sovereign galaxy of about a hundred billion or so stars and untold numbers of planets. But the galaxies are so far away that existing telescopes, even the 200-inch at Palomar, can't make them out very clearly.

The greatest distance to which we can see galaxies, at the absolute maximum, is half of the lifetime of the universe ago. Further than that they're just too dim to be seen with existing telescopes. But fortunately, the early universe appears to have been inhabited by a class of objects called the quasars. The quasars, which may have been the nuclei of young galaxies going through a violent youthful phase, are so bright, they shine so brilliantly, that we can see them at much greater distances than we can see galaxies. If we can call up an image of a quasar--

This one is so far away that its light has been distorted by a galaxy lying between us and the quasar and the result has been a double image. These two dots are actually of a single quasar; a quasar whose light has been travelling for so long that we see it as it was when the universe was less than five billion years old, back when the universe was less than 1/3 its present size.

The quasars are so bright that thousands of them have been detected with telescopes here on earth, and, in fact, we could see them at even greater distances than we do, if there were any. But at very great distances--getting back toward 15 billion years ago--the Palomar telescope and other large telescopes find no more quasars. The explanation seems to be that we're penetrating back to a time when the universe was so young that there hadn't been an opportunity for stars and galaxies and quasars to get organized out of the primordial material and start shining. So way back here at extremely long times ago, there's an epoch of darkness.

And yet, even before that, it's possible to see another form of energy, the energy left over by the explosion that began the expansion of the universe, by the big bang itself. This energy permeates the universe, but it's been so thinned out by the cosmic expansion, that it shifted down from the wavelengths of visual light into the radio spectrum. And this cosmic background radiation, as it's called, can be detected using a sophisticated radiotelescope, or as chance would have it, by using an ordinary television set.

Any TV set hooked up to an antenna can detect the ancient photons from the cosmic background radiation. To see them, turn down the brightness control and tune the set to an empty channel--not right now, hopefully, but after the show--and about one percent of the specks of snow that you'll see on the screen are photons left over from the big bang itself. They are relics of the infancy of the universe, particles that have been hurtling through space since before the first stars and galaxies were born.

The legacy of the big bang is still with us. The heat released by the sun and the other stars represents a fraction of the energy stored in the nuclei of atoms at the outset of time. It was then, when the universe was still bathed in fire, that nature would have worked in the marvelously simple way glimpsed through the unified theories.

We've seen by looking at nature on the subatomic scale that the structures of matter and energy, and even the nature of the fundamental forces thought to govern their behavior throughout the universe, would have been simpler under conditions of extremely high energy. And we've seen by looking at nature on the large scale--out in the expanding universe of galaxies--that the universe began in just such a state of high energy. By putting these two lines of inquiry together, scientists have been able to trace the broad outlines of cosmic history from the first fraction of a second of the big bang down to the present day. We don't know all the details of this story by any means, of course. Parts of it are missing, much of what's been surmised is doubtless distorted or simply wrong. But even at this early stage it's possible to discern the grandeur and beauty, and the extraordinary explanatory power, of what is after all the ultimate history story--the history of the universe, the story of how a single kernel of energy could have become everything that there is.

To examine the account of genesis and evolution in detail, suppose that the steps leading up to the tower of this old lighthouse could carry us backward into cosmic history, so we could scrutinize every thread in the long tapestry of time. Let's imagine that each of the windows in the lighthouse looked out on an earlier epoch in cosmic history, so that this first window let us see the way the universe looked when it was only one billion years old. And that each step up the stairway,, from there on, took us back to when the universe was one tenth its previous age after the big bang, then 10 million, one million and so forth.

Walking in this way we very soon would have reached the first second of time. And that's important, because a lot happened during that first second. Our galaxy--and pretty much all the other galaxies so far as we can tell--formed during the first billion years of the expansion of the universe when the primordial gas was still thick enough to congeal readily into stars and galaxies. We don't know all the specifics of how the galaxies formed by any means, but we think we have a pretty good picture of what the young Milky Way might have looked like--and here it is.

These first generation stars were composed almost entirely of hydrogen and helium. If there was any intelligent life in the early days of the universe, the periodic table up at the front of high school chemistry class would have had just two squares in it--hydrogen and helium.

Once the universe had been expanding for about one million years, it had thinned out enough so that photons could fly freely through space without constantly running into other particles. The result was the dawn of light. This was also the date of the birth of the first atoms. Free at last from harassment by the photons, electrons could settle down in orbit around atomic nuclei. One electron orbiting one proton gives you an atom of hydrogen, the simplest and most abundant of all the elements. Two electrons orbiting a pair of protons--plus a couple of neutrons--gives you a tidy little atom of helium. But those helium nuclei were already on hand. They must have formed at an even earlier epoch.

The creation of helium nuclei dates from when the universe was about 100 seconds old. That's the first point at which things had cooled off enough so that protons and neutrons could get together and form nuclei of atoms without constantly being torn asunder again in the all pervasive heat.

Now, for the first time, protons and neutrons are able to cling together--thanks to the strong force. They tend to form triplets, and the triplets combine to make an unstable nucleus. Two of the extra protons are thrown off and the result is a stable nucleus of helium.

This elaborate mating ritual seems to have been pretty popular in the early days--so popular that the theorists calculate that about one quarter of all the stuff in the universe should have congealed into helium gas. And sure enough, when astronomers study the chemical composition of the universe at large today, they find that it's about one quarter helium. It's this sort of confirmation of theory by experiment that leads scientists to think that they really do understand something of how helium atoms formed in the fires of the big bang.

We're climbing now into very early times. When the universe was one second old the heat was so intense that it overwhelmed even the strong nuclear force. That's the force that holds quarks together to make protons and neutrons. From here on up, even such fundamental structures as protons and neutrons can't exist, and the universe is a soup of free quarks.

A tenth of a second, 1/100th of a second after the beginning of time--the universe now is so dense that even neutrinos, subatomic particles so aloof that they can normally fly through a trillion mile block of solid lead without hitting anything, even neutrinos are now bound up in the universal broth of matter and energy.

One 10 billionth of a second after the beginning, the heat was sufficiently intense that the electromagnetic and weak forces were still welded together and functioned as a single unified force, the electroweak force. Z particles could be created in abundance out of the heat of the electroweak epoch. Weak bosons and photons acted interchangeably, and the universe was ruled, not by four forces, but by three.

When the universe was just a tiny fraction of a second old--about 10 to the minus 35th of a second to be exact--it's thought that the electroweak force had not yet diverged from the strong force. For one brief, shining moment, the natural simplicity envisioned by the grand unified theories was reality. Exotic X particles and free quarks sailed the subnuclear seas. Gluons, photons, and weak bosons danced interchangeably.

So, only two forces were operating in the extremely early universe--gravitation and this electronuclear, or "grand unified" force, as it's called. Yet it's possible that even earlier in the history of the universe, things were even simpler.

We've reached the first instant of time. The fraction of a second that has elapsed since the universe began to expand is so small that it has no name. To express it you'd have to write a decimal point and then a string of 40 some odd zeros. The universe, everything that there is or can be, was contained, we think, at this point within a single spark of energy rapidly expanding but still smaller than the nucleus of an atom and ruled by a single primordial law.

If we knew what went on at this epoch, we might finally understand the relationship between the laws of nature and between space and time and matter and energy. But we don't yet know. We lack a theory that can explain how nature would have behaved under these extreme circumstances. A lot of people are looking for such a theory. Some think it will be a kind of quantum gravity, or what's called a superunified theory or a supersymmetry theory. And of course, we don't know what the theory will say. But whoever hits upon that theory will be the first to have glimpsed the very threshold of creation.

MICHAEL TURNER: Probably the most fundamental question that we can ask about the universe is, "What got it started, where did it come from, the moment of creation?" And that's probably the most difficult thing to try to answer, because in cosmology the way we reconstruct the history of the universe is to run the movie backwards. The way we run the movie backwards is by using the laws of physics. The laws of physics that we presently know are probably good enough to take us back to within 10 to the minus 43 second of the bang or the moment of creation. That's pretty close. But in order to go all the way back we've got to get a better theory of gravity. We need a quantum theory of gravity. I suspect that we may always find ourselves in this position--that to go that last tiny fraction of a second, we need some knowledge that we don't have. And so I think it may be a very long time, if ever, before we can answer the question that everyone would like to know--What caused creation?

STEPHEN HAWKING: It may be that the universe did not really have a beginning. It may be that spacetime forms a closed surface without an edge, rather like the surface of the earth, but in two more dimensions. If the suggestion that spacetime is finite but unbounded is correct, that the big bang is rather like the North Pole of the earth. To ask what happened before the big bang is a bit like asking, "What happens on the surface of the earth one mile north of the North Pole?" It's a meaningless question.

ALLAN SANDAGE: If there was a creation event, it had to have had a cause. This was Aquinas' whole question, one of the five ways he did the God: If you can find the first effect, you have at least come close to the first cause, and if you found the first cause, that, to him, was God. What do astronomers say? As astronomers you can't say anything--except that here's a miracle that seems almost... supernatural... an event which has come across the horizon, into science, through the big bang. Can you go the other way, back outside the barrier and finally find that answer, "Why is there something to nothing?" No, you cannot, within science. But it still remains an incredible mystery: Why is there something instead of nothing.

The ancient Chinese thought of being as having arisen from its opposite, nonbeing. "Nothingness," they wrote, "produced the universe and space and time." Prehistoric myths portrayed genesis in biological terms... "The primeval God transformed himself into a azure egg." In the Judaeo-Christian tradition, creation involved order arising from chaos, and light from darkness: "In the beginning God created the heaven and the earth... and the earth was without form and void and darkness was upon the face of the deep... and the spirit of God moved upon the face of the waters and God said, 'Let there be light.'"

A cathedral was the medieval equivalent of a giant particle accelerator. France in the 14th century spent a greater proportion of its wealth on cathedrals than the United States in the 1960s spent to send a man to the moon. Like spaceships or particle accelerators, cathedrals pushed existing technology to its limits. This cathedral in Beauvais, France, the tallest in the world, twice collapsed under its own weight and had to be rebuilt.

Religion and science are sometimes depicted as if they were opponents, but science owes a lot to religion. Modern science began with the rediscovery, in the renaissance, of the old Greek idea that nature is rationally intelligible. But science from the beginning incorporated another idea, equally important: that the universe really is a universe, a single system ruled by a single set of laws. Science got that idea from the Judaeo-Christian and Muslim belief in one god.

Let me read you a prayer. "Great is God our Lord. Great is His power and there is not end to His wisdom. Praise Him you heavens, glorify Him sun and moon and you planets, for out of Him through Him and in Him are all things, every perception and every knowledge." That prayer was written in the 17th century, not by a priest but by an astronomer, Johannes Kepler, who discovered the laws that govern the motion of the planets. The founders of modern science--Kepler and Copernicus, Isaac Newton, and even Galileo, for all of his troubles with the Church--were, by and large, profoundly religious men.

I'm not saying that you have to believe in God in order to do science. Atheists and agnostics have won Nobel Prizes, as have Christians and Jews and Hindus and Muslims and Buddhists. But modern scientific research, especially unified theory, testifies to the triumph of the old idea that all creation might be ruled by a single and elegantly beautiful principle. The churchmen of the Middle Ages built their cathedrals out of stone, but they built them to express ideas. Stone can only go so high. But ideas can reach across the universe.

Wilson Hall, headquarters of the Fermilab particle accelerator, was modeled in part on Beauvais Cathedral. Robert Wilson, the physicist and sculptor who built Fermilab, said he was impressed by what he called the "curious similarities" between cathedrals and accelerators--the way a cathedral achieved soaring heights in space and an accelerator achieves unprecedented heights in energy, and the way both embody what Wilson called "an ultimate expression."

But whatever the similarities between a Fermilab and a Beauvais Cathedral, there are also profound differences between them, and between science and religion. Scientific theories are held hostage to experimental results. Fermilab is being souped up to even higher energies, to submit the new theories to even more stringent trial by ordeal. The men and women who built Fermilab share a scientific belief that the universe is rationally intelligible, and they built this machine to test their faith.

"Broken Symmetry," a Robert Wilson sculpture that stands at the entrance to Fermilab, expresses one of the beliefs of modern physics--that the universe may have begun in a state of perfect symmetry. The theories say that matter froze out of energy while the early universe was expanding and cooling; that form arose from formlessness like ice crystals congealing in a freezing pond. The mathematical symmetries that the unified theories have exposed at the foundations of natural law are more subtle and complex than those of snowflakes, but their principle is the same--they imply that we live in a crystallized universe of broken symmetries.

Perfect symmetry may be beautiful, but it's also sterile. Perfectly symmetrical space means nothingness. As soon as you introduce an object into that space you break the symmetry, by creating a sense of location: there's a place where the object is and another place where it isn't, and out of that comes tumbling all of the geometry of space as we know it. Perfectly symmetrical time means that nothing can happen. As soon as you have an event, then you break the symmetry, and time begins to flow in a given direction.

We live in a universe that's full of objects and events, and that means that the universe is _imperfect_, that the symmetries in the universe that we live in are broken. It may be that we owe the very origin of our universe to the imperfection of the breaking of the absolute symmetry of absolute emptiness. There's even a theory to this effect. It's called "vacuum genesis," and it suggests that the universe began as a single particle arising from an absolute vacuum.

Curious as it may seem, this idea violates none of the known laws of physics. We've seen how virtual particles come into existence all the time from a vacuum and then fall back into non-existence. There appears to be no upper limit on the size or longevity of the particles that can be created in this way. It's just possible that there might have been absolutely nothing out of which came a particle so potent that it could blossom into the entire universe. It's not very likely--but then, it only had to happen once.

The theory of vacuum genesis is a new idea and nobody knows whether or not it's true, but it does satisfy two of the criteria of a sound scientific theory. It seems, at first, so strange that it must be preposterous, and, like the universe itself, the longer you get to know it, the more beautiful it becomes.

Out of nothingness could have come the spark of genesis. As the universe expanded and cooled, darkness descended. Then light dawned anew with the formation of the first stars. Each star is a nuclear furnace where matter is coaxed into releasing a little of the energy it inherited from the primordial fireball. Thanks to imperfection, to the fractured symmetries that produced differences among the particles and forces, atoms in their variety could build themselves into molecules, and molecules rise up in alliance as life, and life give birth to thought, and thought produce theories about the creation of the universe.

JOHN WHEELER: There's nothing deader than an equation. You write an equation down in a square on a tile floor, and on another tile on the floor you write down another equation which you think might be a better description of the universe. And you keep on writing down equations, hoping to get a better and better equation for what the universe is and does. And then when you've worked your way out to the end of the room and have to step out, you wave your wand and tell the equations to fly. Not one of them will put on wings and fly. Yet the universe flies. It's--it has a life to it, that no equation has. And that life to it, is a life with which we are also tied up.

ALLAN SANDAGE: Out of the big bang has come a nonchaotic system--because otherwise cause and effect, which surely exists, would be impossible. So the design that one sees in the universe may be completely natural as an outcome of the differential equations, but the mystery is, why is the world describable in terms of differential equations? And it is. That's the answer physics gives. All students that ever study are mystified by the recipes that the great scientists have found, but the universe works by those recipes. So the universe that we observe is not a chance phenomenon.

MURRAY GELL-MANN: I find that loving nature and working to conserve nature at the level of tropical forests and animals and birds, habitats in general, creatures of the ocean, and so on, is related to being interested in studying the laws of nature, how nature operates, for example at the level of physics, fundamental physics. To me, these are all parts of the same whole. The beauty that nature exhibits if you see some glorious creature in the wild--like the giant river otters in the Cochas of Amazonian Peru--that beauty to me is related to the beauty that we see when we study the fundamental laws of physics. All of these are different ways in which nature shows its beauty.

STEPHEN HAWKING: From the age of thirteen or fourteen, I wanted to know how the universe worked and why--and why it is what it is. But now I have some idea of how the universe works, but I still do not really understand why.

Einstein used to say that the most incomprehensible thing about the universe is that it's comprehensible. It's an old riddle: What is it about the human mind that so resonates with the rest of the universe that we're able to understand anything about the workings of nature on the largest scale?

It's a philosophical question I suppose, but science has been able to provide us with a little bit of the answer. When we look at subatomic particles and when we look at the stars and galaxies, we see evidence from every direction that the universe is all of a piece, and that it began as a single seed smaller than an atom.

And, in a very real sense, you and I were there. Every scrap of matter and energy in our blood and bones, and in the synapses of our thoughts, can trace its lineage back to the origin of the universe. The natural laws that fragmented and multiplied as the young universe expanded and cooled continue to operate today, in the beating of our hearts as well as in the trajectories of the stars.

As the Koran puts it, "The universe is as close as the veins in our necks." The evolution of the universe goes on, not just around us, but within us. Our thoughts and feelings, after all, are part of the universe, too. Its story is our story as well.

****

This transcript was reprinted with the permission of Timothy Ferris.

OVERHEAD PROJECTION TRANSPARENCIES

Following is the listing of the overhead transparencies available
for use with ASTRONOMY: FROM THE EARTH TO THE UNIVERSE, 3rd ed.
Transparencies are distributed through the publisher's local field
reps. If you do not know who your rep is, please find out who he
or she is by calling the appropriate regional office of Saunders
College Publishing:

    Eastern: (212) 599 8435
    Central: (312) 323 0205
    Western: (415) 692 6386.

Experience has shown that direct contact with the field rep is all
but necessary to obtain the overheads.

In making up the set of 100 overheads, I have concentrated on
including illustrations and photographs that are not readily
available in color slide sets.

Transparency Number	Figure Numbers ETU3	Description
1.	page 1	Sense of time
2.	2-11, 2-12	Copernicus's diagram
	2-15	Retrograde motion of Mars
3.	2-16	Tycho's observatory
	Chapter 3 Opener	Galileo's _Dialogue_ frontispiece
	3-14	Newton's velocity of projectiles
4.	3-10	Galileo drawing of the Moon
	3-13	Phases of Venus
5.	4-2	Parts of the spectrum
	4-2	Windows of transparency
6.	4-4	Prism
	4-5	Emission and absorption lines
	4-6	Resolving point sources
7.	4-7	Refraction of a wavefront
	4-8	Refraction
	4-9	Focal length
	4-10	Refracting telescope
8.	4-13	Chromatic aberration
	4-15	Spherical aberration
	4-16	Parallel light
9.	4-17	Parallel light
	4-19	Newtonian telescope
	4-20	Cassegrain, Gregorian telescopes
10.	4-22	Schmidt telescope
	4-24	Sky Survey red and blue photos
11.	4-26	Soviet 6-m telescope
	4-29	Mauna Kea telescopes
12.	4-31	New technology telescopes
	4-32	Hubble Space Telescope
	4-33	Space Telescope mirror
13.	4-36	Light pollution
	4-41	Exposure time effect on Orion
14.	4-38	Doppler effect explained
	4-39	Doppler effect illustrated
15.	4-45A, B	Einstein Observatory telescope
	4-46	ROSAT
	4-48	Kuiper Airborne Observatory
16.	5-1	Constellations

OVERHEAD PROJECTION TRANSPARENCIES

OVERHEAD PROJECTION TRANSPARENCIES

OVERHEAD PROJECTION TRANSPARENCIES

57.	21-19	Giant flare in H-alpha
58.	21-25	Measurements of solar constant
	21-17	Warping of space
59.	22-1	Track of protostars
	Chapter 22 Opener	Herbig-Haro objects
60.	22-2	T Tauri
	22-3	C-S Star
61.	22-5	Hydrogen and helium ions
	22-6	Hydrogen and helium isotopes
62.	22-9	Carbon-nitrogen-oxygen cycle
63.	22-8	Proton-proton chain
	22-10	Triple-alpha process
64.	23-1	Track to red giant stage
	23-4	Object Gómez
	23-5	Central stars of planetary nebulae
65.	23-2	Size of red giant
	23-6	Size of white dwarf
	23-7	Sirius A and B
	23-12	Roche lobes in a mass exchange
66.	24-1	Track to supergiant stage
	24-4	1979 supernova in M100
	24-5C	1986 supernova in Centaurus A
67.	24-7	Einstein and VLA maps of Cas A
	25-3	Pulsar pulses
	25-5	Pulsar distribution
68.	25-10	Crab pulsar
	25-15	SS433, artist's conception
69.	26-1 to 26-7	Black hole/Cheshire cat
70.	26-6	Rotating black hole
71.	25-16	SS433
	26-8	Black hole, computer calculation
	26-10	Black hole, artist's conception
72.	27-3	Parts of our galaxy
	27-7	IRAS view of galactic center
73.	27-13	Optical drawing of the sky
	27-14	X-ray map of our galaxy
74.	27-10	Infrared map of galaxy center
	27-11	VLA images of Sgr A and "Arc"
	27-12	VLA images of Arc filaments
75.	27-16	EXOSAT x-ray map of galactic plane
	27-18	COS-B gamma-ray map of our galaxy
	27-19	IRAS map of sky
	28-4	Frequency vs. wavelength
76.	28-6	Spin-flip in hydrogen
	28-8	Mapping our galaxy
77.	28-12	21-cm maps of our galaxy
	28-13	Rotation curve of our galaxy
	28-15	Visible view of Bok globule
78.	xviii	The Great Nebula in Orion
79.	28-16	IRAS view of Bok globule
	28-17	Makeup of Orion molecular cloud
	28-20	IRAS scan: Large Magellanic Cloud
80.	28-21	12-m NRAO telescope at Kitt Peak
	29-27	VLA radio array

OVERHEAD PROJECTION TRANSPARENCIES

81.	29-3		de Vaucouleurs galaxy classes
	29-7		Hubble/Sandage galaxy classes
82.	29-8		Toomres's interating galaxies
	29-9		The Antennae
83.	29-11B		Distribution of galaxies
	29-13		Hubble's original diagram
84.	29-14		Galaxies and their spectra
85.	29-12		Galaxies on edges of bubbles
	29-15		Hubble's law
	29-16		Raisin cake expanding
86.	29-19		Cygnus A
	29-20		M87
87.	29-23		Interferometry: positions
	29-29		Giant double-lobed galaxies
88.	29-17		Gallactic cannibalism
	29-30		Head-tail source/Westerbork
	29-31		Head-tail source/VLA
	29-32		Galactic filaments and voids
89.	30-2		Quasar: 3C 273 spectrum
	30-4		Doppler shift for z=0.2, z=1
90.	30-9		Stephan's quintet
	30-12		Galaxies and quasars at same z
91.	30-13, 30-14		Seyfert galaxies
	30-15, 30-16		Resolving quasars
92.	30-17, 30-18		Quasars and interacting galaxies
93.	30-23		Double quasar diagram
	30-24		Double quasar radio map
94.	30-19		Superluminal velocity in 3C 273
	30-20		Superluminal velocity explanation
	30-27		Multiple quasar
95.	30-23		Gravitational lens
	30-25		Double quasar and its cluster
	30-26		Double quasar
96.	31-3		Open and closed universes
	31-6		Hubble diagram
97.	31-4		Geometrical universes
98.	31-11		Penzias and Wilson
	31-12		Horn antenna
	31-13		Black body radiation
	31-14		Background radiation curve
99.	32-2		Current abundences of the elements
	32-4		Nucleosynthesis
100.	32-8		Quarks
	32-14		History of the Universe

LABORATORY EXERCISES

Laboratory exercises can be divided into two types: those that deal with the sky and observing and those that deal with concepts involved in modern astronomical research. Students almost always like straight observing: a view of Saturn, the moon, the Orion Nebula, or sunspots can make the semester worthwhile. A telescope is all that must be provided.

If equipment and personnel can be arranged so that each student can take a photograph of a celestial object--a short exposure of the moon or of a planet or a long exposure of a nebula--then the observing experience is that much more memorable.

Solar observing is less often provided. Sunspots can be viewed either through a special filter over the telescope or with eyepiece protection. Addition of an H-alpha filter to permit observing of prominences or surface features such as supergranulation, filaments, and plages adds a new dimension. The sun changes from day to day and so presents never-ending variety to a student.

If a spectrograph can be provided to demonstrate the solar Fraunhofer spectrum, then the concept of absorption lines becomes more readily understandable. It is sometimes logistically simpler to arrange a daytime lab than nighttime observing.

If a planetarium is available, constellations, celestial coordinates, eclipse phenomena, and planetary motions are among the things that can be generated. Sometimes large planetaria will put on a special show more academically oriented than their usual show.

It is more difficult to provide laboratories that explain more fundamental astronomical concepts. Sky and Telescope has published several such interesting labs. Reprints in class-sized quantities can be ordered by the codes given:

LE001   The Moon's Orbit, Owen Gingerich, April 1964, pp. 220-221.
LE002   Spectral Classification, Owen Gingerich, August 1970, pp. 74-76.
LE003   The Rotation of Saturn and Its Rings, Owen Gingerich, November 1964, pp. 278-279.
LE004   Variable Stars in M15, Owen Gingerich, October 1967, pp. 239-242.
LE005   The Earth's Orbital Velocity, Darrel B. Hoff, January 1972, pp. 9-10.
LE006   Proper Motion, Owen Gingerich, February 1975, pp. 96-98.
LE007   Pulsars, Kurtiss J. Gordon, March 1977, pp. 178-180.
LE008   The Crab Nebula, Owen Gingerich, November 1977, pp. 378-382.
LE009   Hubble's Law, Aneurin Evans, April 1978, pp. 299-301.
LE010   Cosmic Distance Scale, Jay M. Pasachoff and Ronald W. Goebel, March 1979, pp. 241-244.
LE011   The Rotation of Mercury, Darrel B. Hoff and Gary Schmidt, September 1979, pp. 219-222.

LABORATORY EXERCISES

LE012   The Orbit of a Visual Binary, Aneurin Evans, September 1980, pp. 195-197.

LE013   Quasars, Darrel B. Hoff, January 1982, pp. 20-21.

LE014   The Rotation of the Sun, Owen Gingerich and Richard Tresch-Fienberg, November 1982, pp. 433-438.

LE015   The Orbit of Mars, Owen Gingerich, October 1983, pp. 300-302.

LE016   The Wilson-Bappu Effect, Donald J. Boyd and V. Marye Ogle, July 1984, pp. 20-23.

LE017   How Far Is the Galactic Center?, Alan Hirshfeld, December 1984, pp. 498-502.

All are available as offprints in class-sized quantities. They can be done in a lab/conference section or can be assigned as homework. Sky and Telescope labs are available from:

Sky Publishing Corporation
49 Bay State Rd.
Cambridge, MA 02238.

Minimum order is $4 for 10 different items or multiples in any combination. Up to 50 items, $.20 apiece; 51-100, $.15; 101-1000, $.10; over 1000, $.08. All items in an order must be shipped to one address.

Ask Sky Publishing Corp. for their catalogue, "Scanning the Skies." They sell many books, star atlases, and display photographs.

* * * * *

A lab exercise on Resolving Power by Haym Kruglak appeared in Mercury, July-August 1982, pp. 133-135.

* * * * *

Several interesting labs have been developed by the Astronomy Program, University of Maryland. Several of these labs are reprinted here with their kind permission:

I   Solar photographs (courtesy of Elske v. P. Smith and Donat G. Wentzel)

II   Mars (courtesy of Donat G. Wentzel and Herbert Frey)

III   Palomar Sky Survey photographs: Stars and Nebulae

IV   Palomar Sky Survey photographs: Galaxies

V   Redshifts

Though copies of the photographs are reproduced here, the labs should be given using original photos. The solar photos and the Mars photos are available for $15.00 and $9.00, respectively, from:

Dr. Donat G. Wentzel
Astronomy Program
University of Maryland
College Park, MD 20742.

* * * * *

LABORATORY EXERCISES

The Royal Observatory, Edinburgh, has also made labs from Schmidt plates. They are available for 35 pounds. The plates are: Asteroids, Comet West, Globular Star Clusters, The Galactic Plane, The Vela Supernova Remnant, The Large Magellanic Cloud, and the Virgo Cluster of Galaxies.

The "Edinburgh Astronomy Teaching Package for Undergraduates," designed by Dr. M. T. Bruck of Edinburgh University, includes eight exercises based on U.K. Schmidt plates and the plates themselves.
    The "Edinburgh Astronomy Educational Package for Schools," on a lower level, includes four packs, each with ten photographs.
    For information, write:
        United Kingdom Schmidt Telescope Unit
        Royal Observatory
        Blackford Hill
        Edinburgh, EH9 3HJ
        Scotland

* * * * *

Laboratory Exercise VI in this Teacher's Guide is a pulsar lab developed by Kurtiss J. Gordon, Hampshire College, Amherst, MA 01002. It is reproduced here with his kind permission. Additional copies are available from him: 1 copy, $.50; 2-10 copies, $.35 each; over 10 copies, $.25 each. A rewritten version of this lab is the seventh Sky and Telescope lab written above.
    Laboratory Exercise VII is a sample observing program.
    Laboratory Exercise VIII deals with magnitudes.

* * * * *

A high-pressure sodium lamp can display emission, absorption, and continuous spectra. (See the note by Haym Kruglak in The Physics Teacher, May 1985, pp. 314-15.) The lamp is from the General Electric Owl Series, Lucalox, 35 W, No. C643 N021, at $75; with a clear bulb it is No. 11345 at $45.

* * * * *

The American Journal of Physics has printed Resource Letter EMAA-2 by Haym Kruglak in their September 1976 issue. This resource letter deals especially with laboratory exercises. It is available for $.75 (not in stamps) together with a stamped and self-addressed envelope from:
        American Association of Physics Teachers
        5110 Roanoke Place
        Suite 101
        College Park, MD   20740

Their EMAA-1 is a resource letter listing additional audio-visual aids and other supplemental materials. An updated list is being prepared.

* * * * *

LABORATORY EXERCISES

A cardboard star finder, accurate enough for use by science majors and simple enough for use by non-science majors, is described in detail in "Cardboard Star Finder," E. L. Holverson, American Journal of Physics, 53 (7), July 1985, pp. 634-637.

* * * * *

An interesting class project would be participation in the Problicom (Projection Blinking Comparison Sky Survey) project directed by:

    Ben Mayer
    1940 Cotner Ave.
    Los Angeles, CA 90025.

Participation should be a potential long-term effort to be worthwhile.

* * * * *

The Physics Teacher contains many articles of interest to astronomy teachers. See, for example, an article on a lab for "Measuring the Moon's Orbit," by E. Jay Sarton, October 1980, pp. 504 ff.

TELESCOPE MANUFACTURERS

Celestron International   (213) 328-9560
2835 Columbia St.   (800) 421-1526
Box 3578
Torrance, CA 90503

Meade Instruments   (714) 556-2291
1675 Toronto Way   (800) 854-7485
Costa Mesa, CA 92626
    Telescopes and eyepieces.

Bushnell Division   (203) 282-0768
Bausch & Lomb
c/o George Atamian, Educational Manager
135 Prestige Park Circle
New Hartford, CT   06108
    Compact Dynamax telescopes; 15-cm (6") Newtonian telescopes.

Unitron, Inc.   (516) 822-4601
175 Express St.
Plainview, NY 11803
    Refractors.

Questar   (215) 862-5277
Box 59
New Hope, PA 18938

Edward R. Byers Co.   (619) 256-2377
29001 West Highway 58
Barstow, CA 92311
    Steady mounts.

LABORATORY EXERCISES

CATALOGUES OF TELESCOPES AND ACCESSORIES

Roger W. Tuthill, Inc.   (201) 232-1786
Box 1086   (800) 223-1063
11 Tanglewood Lane
Mountainside, NJ 07092
   Celestron, sun filters including neutral density and H-alpha,
   dew caps, and eyepieces; many of his own handy accessories.

Edwin Hirsch   (914) 786-3738
168 Lakeview Dr.
Tomkins Cove, NY 10986
   Celestrons.

Kalmbach Publishing Co.   (414) 272-2060
1027 North 7th St.
Milwaukee, WI 53233
   Astroscans and 7 x 50 binoculars.

Optica b/c Co.   (415) 530-1234
4100 MacArthur Blvd.
Oakland, CA 94619
   Accessories and filters.

Orion Telescope Center   (408) 476-8715
PO Box 1158
Santa Cruz, CA 95061
   Telescopes, eyepieces, and telescope parts.

Robert T. Little   (212) 834-1888
P.O. Box E
Brooklyn, NY 11202
   Telescopes and accessories.

DayStar Filter Corp.   (714) 591-4673
Del N. Woods
P.O. Box 1290
1513 E. Third Street
Pomona, CA 91769
   H-alpha filters for the sun, solar heliostats.

A. Jaegers   (516) 599-3167
691 Merrick Rd.
Lynbrook, NY 11563
   Miscellaneous lenses, mounts, and accessories.

Edmund Scientific Company   (609) 547-3488
300 Edscorp Bldg.
Barrington, NJ 08007
   Telescopes and accessories of all types.

LABORATORY EXERCISES

Tele Vue   (914) 735-4044
20 Dexter Plaza
Pearl River, NY 10965
    Eyepieces and some telescopes.

In addition, check the advertisements in <u>Sky and Telescope</u> and <u>Astronomy</u> each month.

* * * * *

AstroMedia Corp., publisher of <u>Astronomy</u>, also publishes a quarterly magazine about making telescopes. Subscriptions are $12/year, $21/2 years. To order, write:
        <u>Telescope Making</u>
        AstroMedia Corp.
        P.O. Box 92788
        Milwaukee, WI 53202.

Their <u>Deep Sky</u> (published quarterly) deals with observing. Subscription: $12/year, $21/2 years, single issues $2.95.

AstroMedia is now a division of Kalmbach Publishing Co., 1027 North 7th Street, Milwaukee, WI 53233, (414) 272-2060.

* * * * *

LABORATORY EXERCISE I

The Sun and Solar Activity

The sun is a common star, but because it is so near to us we can see many details on its surface that are invisible on other stars. Many of these "details" affect our lives, some regularly, some sporadically. In this exercise we will concentrate on photographs of some of the solar "activity," features that change over hours, days, or weeks. Their causes are not well understood. We obtain clues to explanations by looking for features that occur together with others. The "theme" that unifies this exercise is the influence of sunspots on ever higher layers in the solar atmosphere--and indirectly on us. These layers, just to remind you, are the <u>photosphere</u>, the <u>chromosphere</u>, and the <u>corona</u>.

You should receive five sheets of pictures. All are taken on or within a few days of March 7, 1970, when an eclipse of the sun could be observed over most of the U.S. eastern seaboard.

Two of the photographs are sketched below. They show various spectroheliograms, that is, photos taken with light at the wavelengths indicated. The pictures are slightly oval because of the nature of the telescope used. The "tick mark" in each picture indicates north as seen from the earth; it is <u>not</u> the pole of the sun.

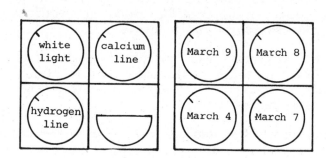

Figure 1   Figure 2

1. First we use the photograph indicated in Figure 1. In the white-light picture you are looking at the layer of the sun called the _____. The most obvious "blemishes" visible on this picture are the _____. For later reference, <u>mark</u> their positions by small dots within the circle on the right (Figure 3).

Figure 3

2. The hydrogen and calcium photographs (spectroheliograms) represent the layer of the sun called the _____. The brighter areas in the hydrogen and calcium pictures are called <u>plages</u>. Are all sunspots surrounded by plages? _____. <u>Sketch</u> the most important plages in Figure 3.

3. Now we want to compare the calcium picture with the x-ray picture. The more nearly white a region is on the x-ray photo, the more x-rays arrive from it. The x-rays come from the layer of the sun called the _____.

Turn your calcium picture so that its pattern matches the pattern on the x-ray picture.

The basic reason why the corona emits x-rays is that it is _____. The fact that the plages and x-ray patterns match suggests that the corona over the plages and thus near the sunspots has a _____ temperature than over other parts of the sun.

4. The x-ray picture was taken immediately after the eclipse. (In fact, the moon still obscures part of the sun's x-rays at the left.) Now we can match the x-ray picture to the picture of the corona taken during the eclipse.

Turn the picture of the corona so that it matches the x-ray picture. Hint: It may be easiest to match the two pictures by matching that place in the corona with no streamer to that place on the rim of the sun on the x-ray picture with the least x-ray emission.

Now use the x-ray picture as an intermediary to match the picture of the corona to the calcium picture. When you have done that, sketch the largest streamer of the corona (the one opposite the place where no streamer appears at all) onto Figure 3, which already contains the sunspots and your sketch of the plages.

Some of the streamers may "attach" to active regions behind the edge of the sun. Is there a streamer that is attached to a region behind the edge on March 7? Check on the hydrogen photos taken on earlier and/or later days.

Review question: Why did we have to wait for the moon to cover the sun's disk in order to take that eclipse picture, but we did not need the moon for the x-ray picture? Why do we not see the photosphere in the x-ray picture?

5. The conclusion of this part of this exercise should be: One way or another, the coronal streamers, the hottest parts of the corona, and plages are ultimately associated with _____.

6. There is one more kind of activity that is related to the sunspots through the spots' magnetism. These are the large arches of gas that appear on the hydrogen picture, as dark filaments in front of the sun's disk and as prominences adjacent to the disk.

Inspect the hydrogen photos of March 4 to 9 (corresponding to Figure 2). Look for evidence that a prominence and a dark filament are indeed the same object, merely recognized differently due to the different position.

LABORATORY EXERCISE I

The best evidence is in the _____ part of the hydrogen picture of March _____ . Describe or sketch the evidence.

7. In the upper left part of the photo of March 4 is a large horseshoe-like filament. Look for its position on subsequent days. Why does it appear in a different position each day? _____.

Draw the sun's equator onto Figure 3, using your best guess for the sun's rotation obtainable from the motion of the filament.

The plages on the hydrogen and calcium photographs (sketched in Fig. 3) appear as bands. How are these bands arranged relative to the equator? _____.

8. Finally, compare the filament in the hydrogen photo of March 7 to the pattern on the magnetogram. The magnetogram measures the magnetic field in the photosphere; what appears black has one magnetic polarity, what appears white has the other. Grey is neutral. We might expect the filament to have some relation to the magnetogram because the filament is also caused in part by magnetic fields. What is the relation between the location of the (one or two) major filaments and the magnetic pattern?

* * * * *

Note that the lab should be carried out with original photographs, which are available for $15.00 per set from:

> Dr. Donat G. Wentzel
> Astronomy Program
> University of Maryland
> College Park, MD 20742.

The lab was developed by Elske v. P. Smith and Donat G. Wentzel.

The photographs are not reproduced here. The following photographs from the text shows the same phenomena but, unlike the ones in the lab, are not all taken on the same day.

	location in ETU, 3rd ed
White light (November 9, 1979)	Ch. 21 opener, page 357
Hydrogen line (November 9, 1979)	Fig. 21-21, page 372
(August 7, 1972)	Fig. 21-19A, page 371
Magnetogram (November 9, 1979)	Fig. 21-14B, page 367
(March 15, 1986)	Fig. 21-14C, Page 367
X-ray picture (June 1, 1973)	Fig. 21-11, page 365
Coronagraph picture (March 8, 1970)	Fig. 21-9, page 364
Eclipse picture (June 11, 1983)	Fig. 21-8, page 363

LABORATORY EXERCISE II

Relative Ages of Various Regions on Mars

Objectives:

i) Identification of the main features on Mars and their relative ages;

ii) Use of crater density and superposition of features to determine relative ages.

One way to learn about past stages in the evolution of the Earth is to investigate the Moon, Mercury, and Mars, which preserve many of the phenomena that happened some 3 to 4 billion years ago and that have long vanished on the Earth.

Ideally, we would like to date craters, volcanoes, etc., absolutely; that is, we would like to determine how many years ago these features were formed. For instance, radioactive dating of lunar samples returned to Earth enabled scientists to determine the approximate time when lava flowed on the Moon (about 4 billion years ago). In practice we have been able to do this only for the Earth and for a very few samples from the Moon.

More generally, photography and mapping of planets and their features can tell us relative ages; that is, they help determine whether a canyon or volcano was made before or after some craters on the same planet.

In this exercise, you are given three maps of Mars, each carefully constructed from many photographs taken by Mariner 9, which went into orbit about Mars in December 1971. We hope that you will become familiar with the main features on Mars while using mainly two methods of relative dating.

A) The first method depends on crater density, that is, the number of craters in any given area. The idea is that rocks of all sizes have been falling on the planets ever since the planets formed, and the impacts produced craters. If the craters in an area are never again altered and are not covered up by lava or dust, then the number of craters is so large that they generally overlap. Large portions of the Moon and Mars are totally covered with overlapping craters. But if the craters are flooded with lava (as on some portions of the Moon and Mars) or eroded (as on Earth) or covered with dust (as for very small craters on Mars), then we see fewer craters in such an area, namely, only those craters that have formed since these destructive events. Therefore, the fewer the craters, the younger is the surface.

1) In the sketch to the right, the surface that is younger is on the (right, left). (Circle answer.)

LABORATORY EXERCISE II

One of the photographs you have is a <u>general</u> map of Mars: only the polar regions are omitted. Later on you will find a <u>sketch map</u> that matches the general map and indicates some of the major features we shall consider: the four major volcanoes, a huge canyon, two of the very large craters, and two even larger fairly circular smooth regions.

2) <u>Identify</u> these features also on the general map.

Roughly half of Mars consists of <u>plains</u>, fairly smooth regions with a few features such as volcanoes or the large canyon superposed, and roughly half consists of regions where the craters generally overlap. Other information tells us that these cratered regions are higher up than the plains, and we call them <u>highlands</u> (as on the Moon). The irregular line from the western to the eastern edge of the sketch map suggests the boundary between the half of Mars consisting of plains and the half consisting mostly of highlands.

3) The region south of the border is the (older, younger) surface.

B) The estimate of which part is older that you must make is based on craters of all sizes visible on the photograph. An <u>independent</u> estimate can be made using the "very large" craters, such as the two indicated on the sketch map (one at latitude S52, longitude 80, which is double-ringed, the other at latitude N50, longitude 330).

1) Are most of the "very large" craters located in the highlands? _____

2) Find at least two "very large" craters that do <u>not</u> lie in the highlands. <u>Indicate</u> them on the sketch map by small circles.

3) Make the following estimate: The highlands contain about _____ times as many "very large" craters as do the plains.

4) <u>Suppose</u> now that we could measure an absolute age for the plains of Mars of 3 billion years <u>as we can do on the Moon</u> and <u>suppose</u> also that the rate of cratering (number of craters formed per billion years) has always been the same in the past. Then, what age would you deduce for the highlands and for all of Mars, using your estimate in part (3)? The age would be roughly _____ billion years.

5) What is wrong with that conclusion? Which supposition is easier to give up?

Scientists deduce from this kind of information for the Moon that most of the large rocks (and asteroids) throughout the solar system collided with each other (making smaller rocks) and with planets (making very large craters) during the first half billion years of the solar system. Therefore the "very large" craters should appear mostly on the oldest surfaces. This should be true not only for the Moon but also for Mercury and Mars.

6) What does the existence of asteroids today and of "very large" craters on the <u>plains</u> of Mars imply about the accuracy of the previous paragraph?

C) Another method of relative dating is the method of <u>superposition</u>: whatever is on top of or partially obliterates another feature is the younger.

1) In the sketch to the right, the smaller crater is (younger, older).

The two very large, fairly circular, and smooth regions indicated on the sketch map were probably also caused by impact, which required a very large rock or small asteroid. However, the smooth surface suggests that this <u>surface</u>, which is a lava flow, was made quite late, after most of the bombardment and cratering on Mars was over. We are left with two choices:
a) These giant impacts occurred early, when most impacts occurred; they excavated a large basin with a rim around it (see sketch), and the lava filled the basin only very much later, or,

b) The giant impacts just happened to occur rather later, after most other impacts, and the lava filled the basin at that time.

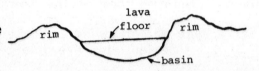

2) Inspect the general map near the two large basins and identify a rim for each, or at least a fairly circular ring-like structure around each. Now choose between choices (a) and (b) above on the basis of the appearance of this rim. The correct choice is
_____ because:

3) Does your choice agree with the earlier conclusion (of Part B) that most "very large" craters were formed rather early in the evolution of Mars? _____

D) You are also given two more detailed maps. They fit next to each other at longitude 90. We shall call the left map "map V" because it contains a volcano and the right map "map C" because it contains most of the giant canyon. (The top of map V fits next to the bottom of map C as reproduced here.)

1) Identify the two maps with the corresponding portions on the general map. Draw the boundaries of the maps V and C onto the sketch map, using the general map for comparison. While you are doing this, notice the change in the scale of the maps and of the sizes of craters that show up; none of the "very large" craters appears on maps V and C. Therefore, the craters that now become important are of modest size. However, they are still large enough so that dust has not covered them up over the history of Mars.

2) All of the regions in maps V and C are younger than the highlands. Explain why this follows from earlier parts of this exercise:

Within this region, we can still distinguish surfaces of at least three different ages.

3) Look at the area south of the great canyon, between longitudes 45 and 100. (You need both maps V and C.) This area is not uniformly cratered (or wrinkled). A boundary seems to occur at about longitude _____. The older surface is to the (right, left).

4) On map V, look at the very smooth plain that extends from the top of the map (longitudes 103 to 110) down to latitude S24, longitude 109. Call it the smooth plain in the remainder of this exercise. The low crater density suggests this smooth plain is (younger, older) than the area south of the main canyon.

5) On map C, check the crater density within the canyon and both north and south of it. In principle the canyon might have been made (1) before the surrounding plains were made, and the lava floods that made the plains simply did not reach the canyon; (2) as a boundary between two plains of very different ages; or (3) after the plains, which would have covered all of map C before the canyon was made. The most correct choice based on your observations of crater density is choice _____ because:

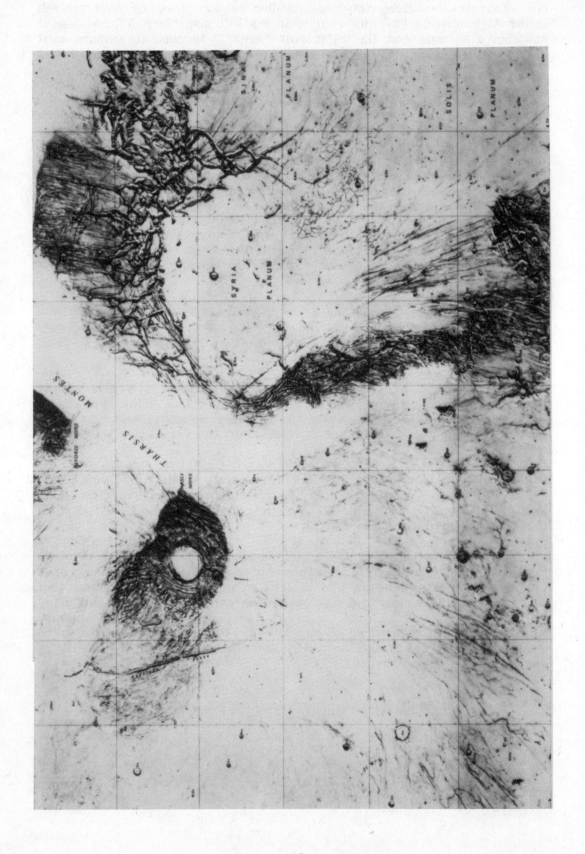

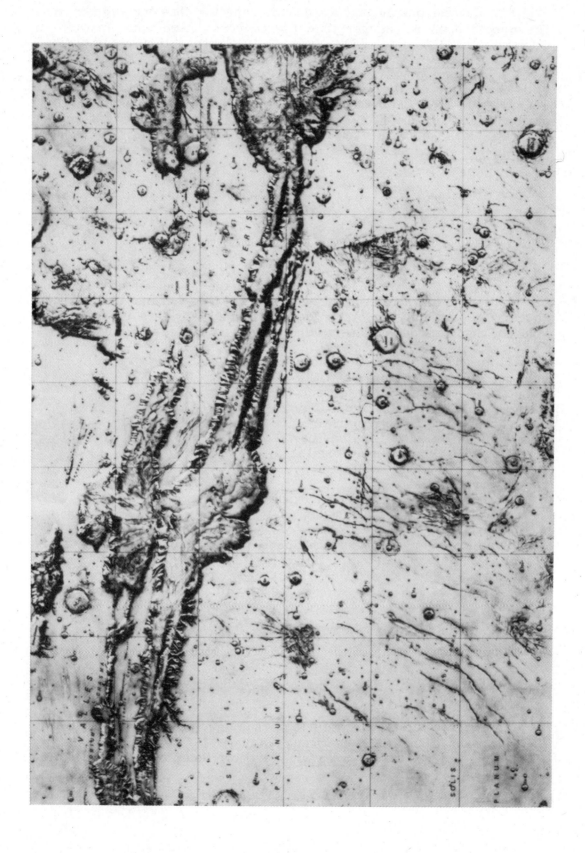

6) Several details that we do not have the time for suggest that the smooth plain is younger than the canyons. Are your answers to questions (4) and (5) consistent with this suggestion? _____.

7) Notice on the general map that each of the four large volcanoes has a crater at its center (that is, on top). Considering the crater density in the areas surrounding the volcanoes, are the craters at the tops of the volcanoes likely to be due to impact? _____. Explain.

8) Return to map V, showing the volcano Arsia Mons. We wish to know whether (1) the volcano grew after the plains already existed, or (2) lava flows created the plain after the volcano was already there. We can not use crater density because craters on or near a volcano might be due to the volcano rather than due to cosmic impacts.

Find several extended structures such as ridges or ditches on Arsia Mons that almost surely were made by the volcano. Do any extend out onto the plains and especially onto the smooth plain? Sketch them on the right. Which choice concerning the ages of the volcano is indicated by the observations?

_____. Explain.

E) Finally, as a summary, indicate the relative ages of the six regions we have dealt with, indicating by 1 the oldest and 6 the youngest:

_____ highlands   _____ smooth plain, north-south
                                          near longitude 10

_____ south of canyon,   _____ Volcano Arsia Mons
      longitudes 45-85

_____ south of canyon,   _____ canyon (Valles Marineris)
      longitudes 85-100

If you are interested in further readings, see the articles by W. K. Hartmann, Scientific American, January 1977, page 84, and by Michael H. Carr, American Scientist, November-December 1980, page 626.

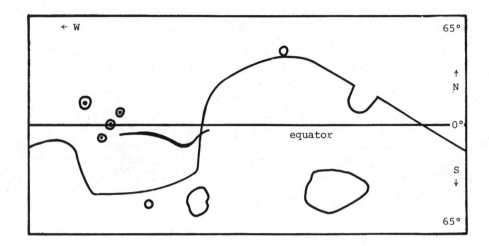

NASA photographs, in part compiled by Herbert Frey. Laboratory exercise by Donat G. Wentzel and Herbert Frey. Photos 1 and 2 are reproduced on the preceding pages. For actual classroom use, originals should be ordered for $9.00 per set from:

Dr. Donat G. Wentzel
Astronomy Program
University of Maryland
College Park, MD 20742.

NOTES FOR LABORATORY EXERCISES III AND IV

Palomar Observatory--National Geographic Society Sky Survey

Description of Prints and Suggestions
for Use in Laboratory Exercises

Dr. Donat G. Wentzel
Astronomy Program, University of Maryland

Two prints with clusters of galaxies and two pairs of prints covering adjacent regions in Cygnus have been selected for use in classroom exercises. They are not the most dramatic of the Palomar prints, because little can be done with very complicated prints beyond superficial viewing. The wide field of the Palomar prints also makes special objects (nearby galaxies, star clusters) less interesting on Palomar prints than on other commercially available photographs.

The brochure mailed with the sets describes the basic features of the Palomar Survey. This includes information on image scale, limiting magnitude, differences between red and blue prints, and a description of several types of defects that may occur. It may be necessary for users to obtain independently clear covering for each print and an eyepiece or magnifier (about 6-power, preferably with scales) for each print or student. Much of the astounding detail on the prints appears only with a magnifier. Backing for the prints and an overlay grid system are also useful.

Laboratory sessions with these prints should be rather loosely structured so that students can make their own discoveries. The sense of discovery is raised when students work in pairs. I acknowledge valuable help by Dr. E. Smith and Dr. T. Matthews for both print selection and the following suggestions.

Nebulae:

E-1099, O-1099. +48° 20^h24, red and blue prints. 1950 coordinates of center: $20^h26^m56^s$, 48°18'56". In Cygnus, including some nebulosity that is centered on the Cygnus X region (next print south, described below) and is reminiscent of the Veil Nebula. By comparing red and blue prints, pick out bright red and bright blue stars. Compare the nebulosity on red and blue prints. An emission nebula is most evident on the red print (Hα emission) but intense emission nebulae appear also on the blue print (because of higher Balmer lines, some forbidden lines like [OII] λ3727, and sometimes scattering). Look for a planetary nebula (a small, nearly circular emission region). Can you find a reflection nebula (visible on the blue but not on the red plate)? Can you determine which stars cause these nebulae? Given a distance of roughly one kpc and a plate scale of 1.12 minutes of arc per mm, estimate the greatest length of an uninterrupted filament and the size of the smallest recognizable structure (or the thickness of the thinnest filament). The planetary nebula is clearly visible on the red but barely detectable on the blue print; its detection on the blue print forms a good exercise in transferring star identifications from one print to the other.

NOTES FOR LABORATORY EXERCISES III AND IV

While much of the blue print has a rather uniform field of faint stars, there is at least one well-defined absorption region, suitable for star counts in and out of the region. How well do different students agree on measuring star sizes? How well can they determine by themselves how big an area to use for star counts? Counting stars in 4 or 5 size intervals can give some estimate about the obscuring cloud. Let students pick their own comparison regions and discuss their reasoning in their writeups. Why have stars near the plate limit disappeared in the blue print, while obscuration is much less noticeable in the red print? (For the low accuracies obtainable with student star counts, we have estimated that star diameters on the blue print of 0.05, 0.10, 0.05, and 0.25 mm correspond very roughly to B magnitudes of 20, 17, 14, and 11, respectively.)

E-754, O-754. +42° 20^h30, red and blue prints. 1950 coordinates of center: $20^h33^m20^s$, 42°19'44". Rift in Cygnus. The very bright star, also on the edge of the previous pair of prints, is α Cyg, Deneb. Near the opposite corner is γ Cyg, surrounded by the γ Cygni Nebulosity. On the eastern edge of the print is the Pelican Nebula (which, on the next print east, is separated by a dust cloud from the North America Nebula). Behind the rift is the radio source complex Cygnus X, on which many of the filaments in the previous pair of prints (and on the other side of the Milky Way) are centered, at least approximately. Distances in the region vary from roughly 0.5 kpc (North America Nebula) to roughly 1.5 kpc (γ Cygni Nebula).

This pair of prints is much more dramatic than the other pair since the emission and absorption regions are much stronger. In fact, it may take the students some time getting used to interpreting black on the print as bright emission, and vice versa. Compare the emission and absorption patterns in different parts of the print. Estimate relative distances of obscuring clouds from relative densities of foreground stars. Describe the shapes of the many small, sharply defined obscurations (elephant trunks, etc.). Which boundaries might be shock fronts? Look for obscurations devoid of stars to the plate limit. There is at least one globule (dust cloud with a size comparable to stellar images, seen in front of an emission region). Look for a star cluster. Beware of ghosts and diffraction effects from the bright stars.

Galaxies:

O-1563. +12° 12^h24 blue prints. 1950 coordinates of center: $12^h28^m48^s$, 11°28'42". Part of the Virgo cluster of galaxies. M87 is above plate center. Many galaxies are readily recognizable as spirals, barred spirals or ellipticals, and some subclassification is possible. Let the students establish a classification scheme based on their own drawings of selected galaxies. Among the spirals, what fraction has smoothly wound arms? Are there peculiarities? dust lanes? Which edge-on galaxies may correspond to a chosen face-on spiral? Distinguish barred and edge-on dwarf galaxies (lower surface brightness and smaller diameter). Within a given area, what are the proportions of ellipticals : spirals : dwarfs? How does one distinguish faint galaxies from stars?

NOTES FOR LABORATORY EXERCISES III AND IV

O-83. +18°16^{h}0 blue print. 1950 coordinates of center: 16^{h}04^{m}17^s, 17°44'28". The Hercules cluster of galaxies. This should be looked at after gaining some experience with the Virgo galaxies since the Hercules galaxies are smaller, harder to recognize even with magnification. The phenomenon of clustering is more noticeable. Is there a well-defined edge to the cluster? Are there subgroupings or pairs in the cluster? How many galaxies belong to the cluster? Compare ratio of spirals to ellipticals with that in the Virgo cluster. Estimate relative distances from the diameters of the five largest galaxies in each cluster. Can you find a ring galaxy in the field of this print? Can you find another cluster of galaxies (in the lower right quadrant)?

The "Set of Six Prints from the Palomar Sky Survey" is available for $31.25 per set of six prints surface postage prepaid from:

California Institute of Technology
Bookstore 1-51
Pasadena, CA 91125.

Add 6.5% tax if ordered from California, or $3.35 extra for foreign orders to prepay airmail. Payment must accompany order.

See Dr. Wentzel's article entitled "All Weather Observing: Student Use of Palomar Sky Survey Prints," in Mercury, January-February 1973, pp. 3-16. He comments especially on the four Cygnus plates.

LABORATORY EXERCISE III

Palomar Sky Survey: Stars and Nebulae

The photographs that we will be using are reproductions of plates taken by the 1.2-m (48-in) Schmidt telescope on Mount Palomar. Schmidt telescopes are designed specifically for photographing relatively large (at least by astronomical standards) areas of the sky with very good definition. They are nearly useless for any other purpose. This particular Schmidt telescope is the largest one in the world and was designed, at least in part, with the idea of compiling an atlas of the entire sky visible from southern California. The atlas took about 10 years to complete (under the auspices of the National Geographic Society and the Palomar Observatory, which is run by the California Institute of Technology) and it has since been invaluable to astronomers. The telescope was large enough that the pictures include the most distant objects known, and yet the field of view was large enough (it is usually quite small for a large telescope) that the entire sky is covered by a reasonable number of photographs. Astronomers use the photographs both for survey work in determining the numbers and kinds of different classes of astronomical objects and for discovering and identifying objects that need to be studied further with other types of telescopes.

The original photographs were made on glass (as are most astronomical photographs) because glass is less subject to the stretching, shrinking, and warping that can occur with the acetate and other bases used for ordinary photographic film. The original photographs are stored in a vault but many copies have been made and sold to various observatories and astronomical institutions around the world. All the copies (ours are prints but transparencies on glass are also available) are _negative_ contact copies because, as a matter of practical experience, these preserve more of the details on the original than do any other types of copies. Each print is about 14" square and covers an area on the sky of 6° × 6°, giving a scale of roughly one degree per two inches. (The full moon would thus be about one inch in diameter.) For each position on the sky, there are two different photographs, one taken originally in blue light and one taken in red light. This lets us estimate the colors of different objects and even, in extreme cases, see objects in one color that are nearly or totally invisible in the other.

These prints are of extremely high quality and are the same ones that astronomers use. They are very difficult to replace so please be extremely careful. NO SMOKING!! NO PENS OR PENCILS ANYWHERE NEAR THE PHOTOGRAPHS!!! DO NOT WRITE ON PAPER THAT IS ON TOP OF THE PHOTOGRAPHS.

Basic Data

In the upper left-hand corner of each photograph (which corresponds to the northeast corner on the sky) is a block containing the basic information about the photograph. This information includes the plate sensitivity (whether it was sensitive to blue light or to red light), plate number (which is of no interest to you except that the red and blue photographs of the same piece of sky will have the same number), the date on which the original photograph was taken, and the astronomical coordinates (right ascension and declination, which are analagous to latitude and longitude on the earth) which indicate the exact position in the sky of the center of the photograph.

Example:

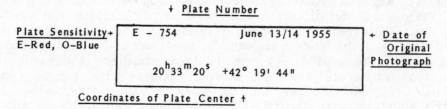

Locating Objects on the Photographs

To locate objects on the photographs, we must have a standardized system. The following system is easy to use and will be the standard in this class. Divide the print into thirds both horizontally and vertically and number the resulting areas as

1	2	3
4	5	6
7	8	9

Then take each of these nine areas and divide them into thirds and number them as above. That gives you the grid system shown below:

1-1	1-2	1-3	2-1			3-1		
1-4	1-5	1-6		(A)				
1-7	1-8	1-9						
4-1			5-1			6-1	6-2	6-3
								(B)
7-1			8-1			9-1		
7-4	7-5	7-6						
						9-7	9-8	9-9

Thus an object can be located within one square by giving its numbers, e.g., A is in 2-5; B is in 6-6. The block of basic information about the photograph is in 1-1. A more complete description would include the plate identification and the location of the object within its square:

      E-1066, 2-5 right side   for A
      O-1066, 6-5 lower right   for B.

You can estimate by eye which square a particular object is in or can make overlays out of tracing paper.

PROCEDURE

Locate the following objects on both red and blue prints and, using the preceding discussion, identify what the type of object is. Look for such things as the sharpness of the edge of the image, the color, the presence of blue stars or a cluster, the overall shape. Be sure to explain _why_ you identify objects the way you do.

1. Identify and give the location of one very blue _and_ one very red star.

2. Identify the white speck on E and O-754, 9-4.

3. Identify the small roundish black object seen on E-1099, 8-1. Look for it carefully on the blue print by locating nearby stars.

4. Identify the three localized diffuse regions on O-754, 6-5.

5. Locate and examine the four emission regions and their associated dust clouds:
      A. E and O-754, 9-4, 9-5, 9-9.
      B. E and O-754, 1-4 to 1-7 and 4-1 to 4-4. This is the westernmost third of an H II region.
      C. E and O-1099, 7-3, 7-6, 8-1, 8-4.
      D. E and O-1099, 8-7.

From a comparison of their size, structural details, amount of dust associated, brightness (blackness), etc., answer the following questions:
      A. Put the regions (A, B, C, D) in order of age, giving your reasoning.
      B. Put the regions in order of distance, again giving your reasoning.

6. Compare the absorbing clouds in E-754, 7-1 to 7-2 and E-754, 9-4 to 9-5. Which cloud is farther away? What is your reasoning?

LABORATORY EXERCISE III

OPTIONAL EXTRAS

7. Look at E-1099 and E-754 together and notice how the filamentary structures tend to curve and suggest that they may be part of a circular structure with its center on the lower portion of E-754. Very near the center is a group of faint stars of spectral types O and B. This group of stars is known as the OB Association Cygnus OB2 and is more strongly reddened than almost any other object that has been measured. If there were no interstellar dust between us and the association, some of the stars would be among the brightest stars visible in the sky. Can you locate this association? (It is near the top of 8-2.) It is also interesting because there is a source of x-rays in the same direction as well as a large, strong, source of radio waves.

8. Examine the boundary between the emission and absorption regions on O- and E-754 at 1-4. This is probably a shock front or "sonic boom."

9. Examine anything else that looks interesting and see what you can deduce about it from a comparison of the two photos or from a comparison with other nearby regions.

This laboratory was developed at the Astronomy Program, University of Maryland.

The Palomar Sky Survey Photographs are copyright © by the National Geographic Society--Palomar Observatory Sky Survey. They are reproduced by permission from the Palomar Observatory, California Institute of Technology.

LABORATORY EXERCISE III

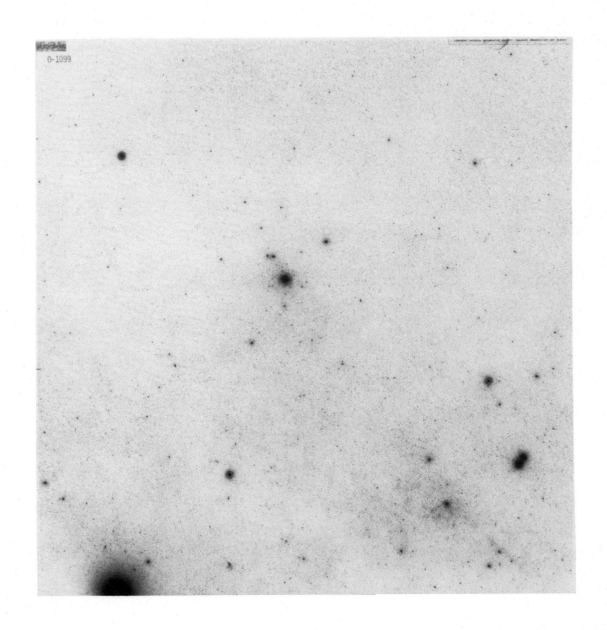

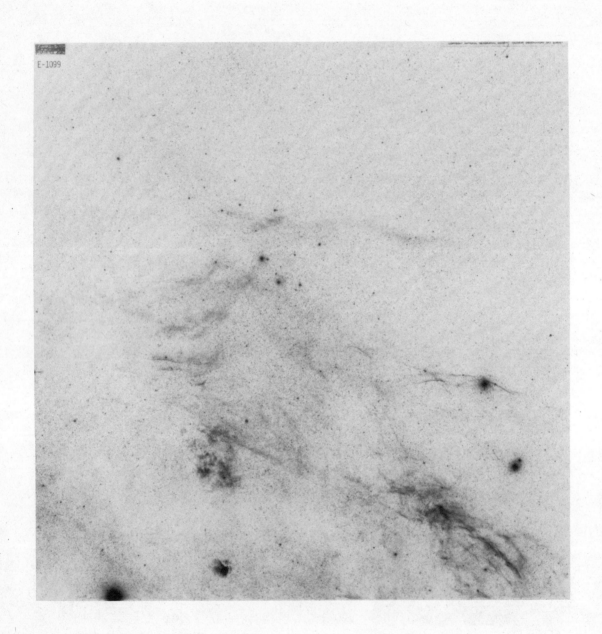

E-1099

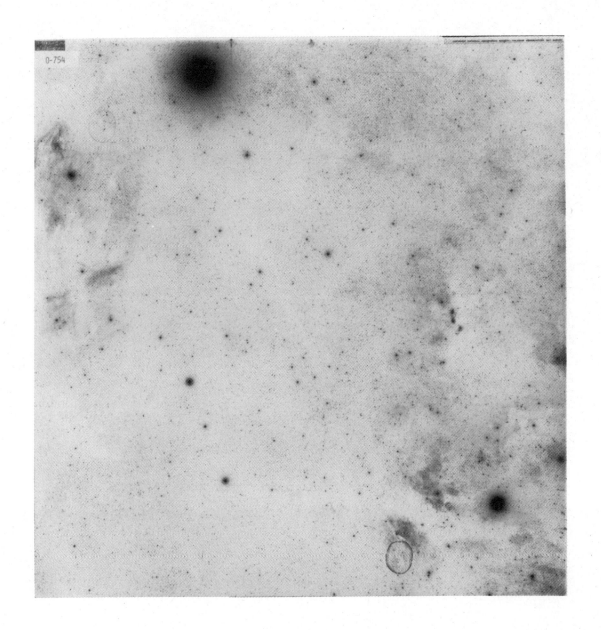

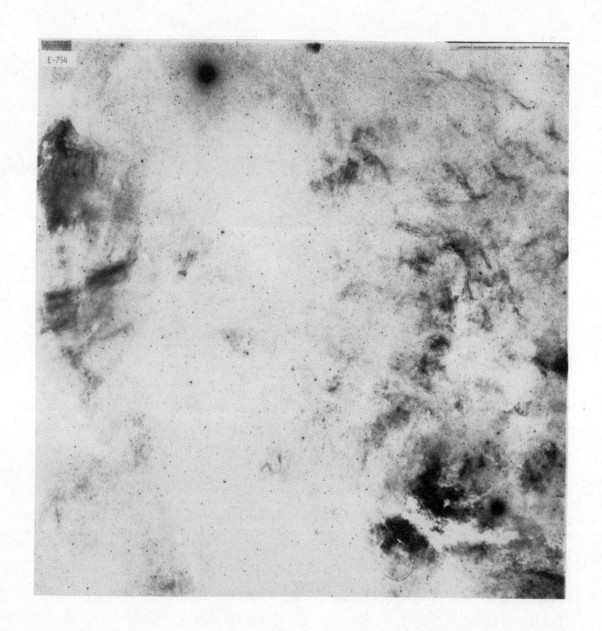

© copyright by the National Geographic Society--
Palomar Observatory Sky Survey

LABORATORY EXERCISE III

Palomar Sky Survey: Stars and Nebulae

Work Sheet for Palomar Prints:
A Milky Way Field in Cygnus

The following instructions and questions are a guide to help you interpret the kinds of photgraphs that astronomers actually use. You should spend much of your time looking at the photographs, because it takes some time to become familiar with them and to sort out the many kinds of information contained in them. The instructor will check your progress and is available to answer questions.

1. The upper left corner of each photograph has a number. Arrange the photographs in two pairs. The pair of E prints (E-1099 and E-754) cover a region in the sky 6° by 12°, in the constellation Cygnus, where we are looking along a spiral arm of our galaxy. The very bright star Deneb is at the line of overlap, as shown in the diagram on the right.

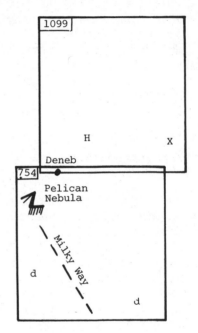

2. The difference between the pairs is the kind of film used. One of the pairs is made from red-sensitive film: it shows as black those parts of the sky that would appear as red in color photographs. The other pair is made from blue-sensitive film: it shows as black those parts of the sky that emit much blue light.

To tell which set is which color, remember first the color of nebulae of hot hydrogen (H II regions):

H II regions emit mostly (red, blue) light. (Circle answer.)

Now compare the two pairs of prints. One set shows many black irregularly shaped regions. These can only be H II regions. That identifies them as the (red, blue) prints. Therefore:

The E-prints are the _____-sensitive prints.
The O-prints are the _____-sensitive prints.

3. Fact: The brighter a <u>star</u> is in the sky, the <u>larger</u> is its image on the photograph. Therefore:

The image of a blue star should be (larger, smaller) on the blue prints than on the red prints.

LABORATORY EXERCISE III

Near the lower right part of Photo 1099 (see position X in the diagram on the previous page), there are two fairly bright stars that appear near each other in the sky. Of the two stars, the more nearly blue star is the (upper right, lower left) star.

4. HERE IS THE MAIN PART OF THIS EXERCISE. Look for the following three nebulae. Each may take some searching. Be careful to look on the correct photograph. The instructor will check that you have found the correct objects.

A planetary nebula is on print 1099. Since it consists of hot hydrogen ejected by a dying star, its color should be (red, blue). Look on the print of the appropriate color. When you have found it, indicate its position in the diagram by a small letter p. Clue: It is small, round, with a sharp boundary.

A globule is a very tiny dust cloud, so tiny that it may soon collapse to form a new star.

Since dust absorbs all light emitted by more distant stars or nebulae, a dust cloud appears (white, black) on the photos.

Therefore, it is best to search for globules in one of the very (white, black) regions of the photos. This suggests that you look on Photo E-574.

Search for the tiniest dust cloud that you can find. Indicate it by a letter g in the diagram of the previous page, and also show it to the instructor.

The globule may look like a speck of dust on the photo. You may be tempted to wipe it away, but it won't go away. How can you check that it is a real globule and not merely a flaw in the film?

A reflection nebula is in the right half of Photo 754. A reflection nebula occurs when dust turns light from a nearby star toward us. The color of a reflection nebula is (red, blue).

Search for the reflection nebula and indicate it by a letter r in the diagram. This takes some patience.

5. An H in the diagram of Photo 1099 marks a fairly large and fuzzy H II region. Find it on the red photo, then look in the same part of the sky on the blue photo. What is there?
_____.

Explain how this object is related to the H II region.

6. The diagram shows where the Milky Way is located, approximately. Now look on the red prints and compare the number of stars (per square inch) in the Milky Way with the number in the upper right part of Photo 1099.

One sees (more, fewer) stars in the Milky Way than some distance away from it.

But supposedly the Milky Way is a disk of billions of stars, and most of these are situated in the direction of the Milky Way. Why, then, do we not see the greatest number of stars in that direction?

The two dust clouds on Photo 754 are indicated by the letter d in the diagram. Each is a thick, opaque cloud. Given this information, how can one tell which of the two clouds is the nearer one?

The (left, right) cloud is the nearer one.

7. Other things to do if you have time:

Look for "ghost" images of stars. They are made when light from a bright star bounces off the photograph, then bounces off something in the telescope, and finally returns somewhere else on the photograph.

Find the planetary nebula on the blue photograph.

Look at the long curved filaments on E-1099 and estimate where on E-754 the center is, approximately. Although it is hard to see on the photo, that center has a large number of hot blue stars, and a super-nova there has left a radio source and an x-ray source.

Imagine trying to give a name to each star in the upper right part of Photo 1099.

LABORATORY EXERCISE IV

Palomar Sky Survey: Galaxies
Version 1

Clusters of Galaxies

All of the large bright galaxies on O-1563 belong to a group or cluster of galaxies called the Virgo Cluster. It is in fact the nearest rich cluster of galaxies. Thus we can say that these galaxies are all at the same distance from us (about 45,000,000 light years) and, therefore, any differences we find in the size or brightness between different galaxies are reflections of the intrinsic properties of these galaxies and are not due to differences in their distance from us.

A more distant cluster of galaxies, the Hercules Cluster, is in the center of the second print O-83. In addition, a very distant cluster is also on this print. This allows us to compare the appearance of galaxies at different distances and to arrive at techniques to estimate relative distances.

Classification of Galaxies

The first thing that a scientist does when confronted with an unknown group of objects is to sort them into similar categories, that is, to classify them. Any classification scheme begins with the most obvious attributes and then considers more and more subtle ones.

1. Examine galaxies number 1-16, all on O-1563. Separate the galaxies into as many different types as you feel are necessary and describe the properties peculiar to each group.

2. We are interested in knowing what the three-dimensional form of the different types of galaxies is like. For instance, a plate looks round when seen straight-on, but looks almost like a line when seen edge-on. On the other hand, a ball always looks round.

In a classification scheme we must be careful not to confuse two different aspects of the same type for two different classifications. (A plate is still a plate, whether seen edge-on or straight-on.)

Examine galaxies 1, 7, 8, 15, and 16. Do all these fall in the same general category or do you need several classes? Explain. What about galaxies 5, 9, 10-14? What would be the general three-dimensional shape for each general class?

3. Note the white streak curving across the nucleus of the edge-on galaxy in 3-4. What do you think is the cause of this white streak? Remember that a white region corresponds to no light from that part of the galaxy.

4. Look at the following 4 objects: 1, 7, 8, and 23. They all belong to the same cluster of galaxies and are thus at the same distance from us. Thus any differences we see between these galaxies,

as a function of their total brightness, are real effects which change with the amount of matter in a galaxy. In particular, note how the surface brightness (brightness per square inch), strength of the spiral arms, diameter, predominance of the nucleus, etc., all change with the total brightness of the galaxy.

   a. Describe the relationship between the luminosity (total brightness) and the appearance of the objects, 1, 7, 8, and 23.

   b. Do objects 24 and 25 (two faint smudges) fit into a logical extension of your scheme? Explain.

   c. In what way do you think the relationship in (a) would be useful? Explain. (There are at least two important ways --try to get at least one.)

5. Examine galaxies 17-20 very carefully.

   a. Do they fit into your classification scheme?

   b. What is their three-dimensional form?

6. On print O-83 examine the cluster of galaxies near the center of the plate (2-8, 5-2, 5-5, 5-6). This is the Hercules Cluster.

   a. What types of galaxies occur in this cluster?

   b. Describe <u>two</u> methods that you could use to determine the relative distances of the Virgo Cluster (O-1563) and the Hercules Cluster (O-83). (There are, in fact, more than two possibilities.) Be sure to state any assumptions necessary for your method to work.

   c. Estimate how many times farther away the Hercules Cluster is than the Virgo Cluster. (Show how you made your estimate.)

   d. If the distance to the Virgo Cluster is 45 million light years, what is the distance to the Hercules Cluster?

7. Examine the faint cluster of O-83, 8-3 to 9-1.

   a. What kind of galaxies does it have?

   b. How may times farther away is it than the Hercules Cluster? Show your method.

   c. Using the answer to question 6d, what is the distance to this cluster?

   d. This cluster is actually one of the most distant objects known and is thus representative of a much earlier stage of the universe. Cosmologists, theerefore, would like to know if galaxies in this cluster are different from galaxies in nearby clusters. Comparing the photographs of the cluster with that of the Virgo Cluster, what selection effects do you think might be present in looking at the photograph of the distant cluster? (Consider, for example, the galaxies mentioned in question 4b. Would you see them if they were in this cluster?)

8. Look at the highly distorted galaxy on O-1563, #22.

   a. Why do you think the galaxy looks as it does?

   b. Give the location of three other very close pairs of galaxies.

LABORATORY EXERCISE IV

Palomar Sky Survey: Galaxies
Version 2

Objectives:

    i) Visually recognize spiral and elliptical galaxies in both face-on and edge-on orientations.

    ii) Estimate the distance to one cluster of galaxies given the distance to another.

    iii) Recognize the importance of practice in looking at photographs of astronomical objects.

Step 1

    Inspect the photograph numbered O-83 (in the upper left corner) for a minute or two. Most of the dots in the picture are common stars in our Milky Way. Look for dots with prongs or small circles around them. These also are stars. The prongs and circles are made by the telescope and camera. This photograph also shows hundreds of galaxies, which are not immediately apparent until you have achieved some experience with the other photographs.

Step 2

    Now that you know what stars look like, study the other photograph, labelled O-1563. The galaxies you see here belong to a cluster of galaxies in the constellation Virgo, so it is called the Virgo Cluster of Galaxies.

    We want you to study the photograph with no magnifiers. Look long and hard enough until you can distinguish:
1. elliptical galaxies (they show no structure, merely get fainter from the center out) from spiral galaxies
2. spiral arms of spiral galaxies that are smooth bands of light from those that are clumpy
3. peculiar things like pairs of galaxies that might be colliding.

    The best way to really look at galaxies is to try to sketch a few. Sketch as great a variety of galaxies as you can find, one each in the five boxes. You do not have to be an artist.

    THESE PHOTOGRAPHS ARE EXPENSIVE AND EASILY DAMAGED! NO SMOKING! DO NOT WRITE ON PAPER THAT IS ON TOP OF THE PHOTOGRAPHS because the impression comes through and appears on the photograph.

These photographs are printed as negatives: stars show up dark; the brighter a galaxy, the darker it appears on the print; conversely, dark sky shows up as white on the print.

LABORATORY EXERCISE IV

Sketch a galaxy in each box:

Step 3

Near the upper right corner of O-1563, just above the giant elliptical galaxy, is an elongated object that looks a bit like a flying saucer because of a white lane through it. Sketch it in the box on the right.

Question: Why are no stars visible where the white lane is? Are there no stars there? Or is something blocking the starlight? Use your knowledge of the effect of dirty city air.

Fact: White lanes like this happen only in spiral galaxies. This is how we know that this object is a spiral galaxy, seen edge-on. Most other cigar-shaped galaxies in the photograph are edge-on spirals.

Step 4

Now return to the photograph O-83. With your new experience you will be able to find a group of several hundred galaxies, clumped in part of this photograph. In the box on the right, sketch the location and outline of this cluster of galaxies (not the individual galaxies). This portion of the sky is in the constellation Hercules, so this is called the Hercules Cluster of Galaxies.

Step 5

Use a magnifier to check whether the Hercules Cluster contains spiral and elliptical galaxies much like the Virgo Cluster does.

Are the galaxies similar? _____

Astronomers assume that the ten largest galaxies in each cluster are in fact very similar in size.

The two photographs each cover a portion of the sky that is 6 degrees by 6 degrees. An object like the full Moon would have the size of a quarter dollar on the photographs.

LABORATORY EXERCISE IV

Why do the galaxies in the Hercules Cluster look so much smaller than those in the Virgo Cluster?

_____ .

Estimate the distance of the Hercules Cluster, given that the Virgo Cluster is 70 million light years away:

_____ .

Hercules Cluster galaxies are typically _____ times smaller than Virgo galaxies. Therefore, the Hercules Cluster is _____ million light years from us.

Step 6

Find at least one more (still further) cluster of galaxies in the lower right-hand quarter of photograph O-83. Indicate its location in the box of Step 4.

Keep this worksheet for future reference.

* * * * *

LABORATORY EXERCISE V

Redshifts

Note: An interesting lab can be done with redshifts of galaxies, using the data from the figure in the chapter on galaxies in the text. Glossy prints of this photograph can be obtained by requesting Palomar Observatory Catalogue #313 ("Redshift: Relation between redshift and distance of galaxies, direct and spectrum photographs shown") from:

> Dorothy Cantrall
> Caltech Bookstore, 1-51
> Pasadena, CA 91125.

The current cost is $9.00 per print.

The following suggestions are based on the lab exercises developed at the Astronomy Program, University of Maryland. They can be used in conjunction with a summary of methods of measuring distances that was assembled by Donat G. Wentzel and R. Hanisch, and that is reprinted following the lab.

1. Measure the shift (in mm) of the H line from its normal position on each of the spectra. The H line is the right-hand absorption line of the part of dark absorption lines marked H+K. The amount of redshift is shown by an arrow on each spectrum.

2. We note from the laboratory that the leftmost and rightmost of the comparison lines are separated by 1130 A. Compute the plate scale, that is, the number of Angstroms per millimeter by which the spectra are spread out. (You can make a rough check that the plate scale you have computed is correct by verifying that the H and K lines are separated by 35 A.) Note that it is much more accurate to use lines widely separated in wavelength to compute the plate scale than it would be to use the H and K lines.

3.   Use the Doppler formula, $v = c\Delta\lambda/\lambda$ to find the velocity of recession of each galaxy.

4.   Graph these results.   You can also include the following additional points:

Cluster of Galaxies	Distance (10^6 l.y.)	Velocity of Recession (km/sec)
Perseus	350	5,400
Hercules	650	10,000
Ursa Major II	2,700	41,000
Milky Way (ourselves)	0 (we are inside)	0

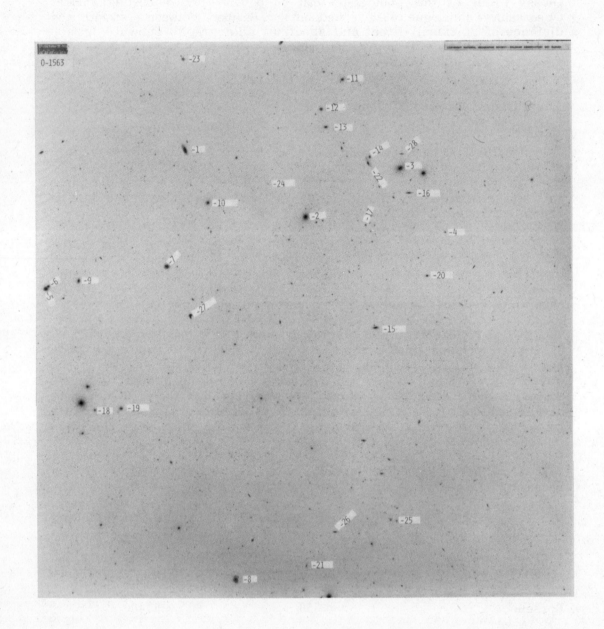

0-83

LABORATORY EXERCISE V

5. Calculate the slope of the line by picking a point on the line and dividing the velocity by the distance. Keep track of the units. The quantity is known as the <u>Hubble constant</u> and is conveniently expressed in units of km/sec/million light years. In these units its value is thus the velocity of recession of a galaxy at one million light years. (If you ended up with km/sec/l.y., you can multiply by 10^6 to get km/sec/million l.y.) The Hubble constant is usually expressed in km/sec/Mpc. What is it in those units? [Note: Palomar Observatory scientists now use a figure of $H_o = 50$ km/sec/Mpc on their photograph. The distances quoted were not independently measured, as they would have been to establish the relationship.]

6. If the distances of the galaxies above were measured by direct methods (Cepheid variables, sizes of H II regions, etc.), then we have established the Hubble relation over this range of redshifts and distances. We can now use the relation to calculate the distances to galaxies that are farther away, if we assume that the Hubble relation still holds. For example, the quasar 3C48 is receding from us at a redshift of 37% of the speed of light. Calculate its velocity of recession, assuming that the non-relativistic Doppler formula holds. How far away is it from us?

7. Until recently, it was thought that Hubble's constant was 100 km/sec/Mpc. How far away would 3C48 be if that were true? What did the change of our knowledge of Hubble's constant do to our idea of the scale of the universe?

8. Find the distance to the Coma Cluster of galaxies, whose velocity is 7,000 km/sec. Find the distance to the Gemini Cluster, which is receding at 23,000 km/sec.

9. How fast is a galaxy receding if it is at a distance of 2 million l.y.? Such a galaxy is so close that it would be a member of our own local group of galaxies. For these nearby galaxies the random motion of the galaxies masks the cosmological redshift.

10. Calculate the age of the universe by taking $1/H_o$ in units of sec^{-1}. Put the age of the universe you calculate--the "Hubble time"--in units of billions of years.

LABORATORY EXERCISE V

Summary: Distance Measuring Methods

METHOD

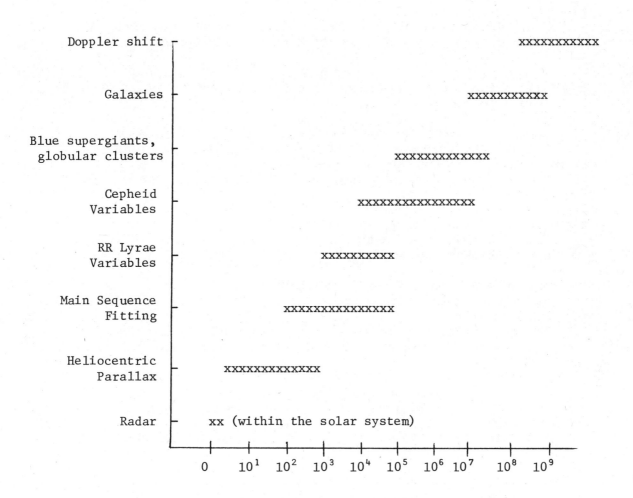

DISTANCE (light years)

The numbers in parentheses indicate the maximum distance that can be determined with the particular method.

Courtesy of Donat G. Wentzel and R. Hanisch, Astronomy Program, University of Maryland.

LABORATORY EXERCISE V

THE DISTANCE SCALE IN ASTRONOMY

<u>Geometrical</u>:

1) Astronomical Unit (mean distance sun-earth)
Found by radar to planets and sun plus Kepler's laws. (Known extremely accurately.)

2) Distance to nearest stars (only to about 100 l.y.)
Found by trigonometric parallax. (Typical error 10%.)
Learned: what types of stars there are; H-R diagram.

<u>Headlight</u> <u>Method</u>: learn luminosity L; observe apparent brightness B; get distance D from $B = L/D^2$.

3) Distance to nearest star clusters (usually within 10,000 l.y.)
Found by recognizing main-sequence stars, thus learning each star's luminosity. (Typical error 10%.)
Learned: RR Lyrae stars have constant luminosity, Cepheids have period-luminosity relation.

4) Distance to galactic center (30,000 l.y.)
Found by distance of swarm of globular clusters, using RR Lyrae variables. (Probable error 10%.)

5) Distance to nearest galaxies (a few million l.y.)
Found using the period-luminosity relation of the Cepheids. (Typical error 10%.)
Learned: brightest objects in each galaxy all have roughly the same luminosity (H II regions, stars, clusters, exploding stars).

6) Distance to nearest clusters of galaxies (30 million l.y.)
Found using brightest objects in galaxies and their (poorly known) luminosities. (Typical error over 50% for any one galaxy, 20% for several in a cluster.)
Learned: brightest galaxies in each cluster of galaxies have about the same luminosity.

7) Distance to further clusters of galaxies
Found by using known luminosity of brightest galaxies in each cluster. (Typical error 20%?)
Learned: Hubble's law; Doppler redshift increased in proportion to distance.

<u>Doppler</u> <u>Shift</u>:

8) Distance to furthest clusters of galaxies and to quasars (10 billion l.y.)
Found by taking spectra, using Doppler shift and Hubble's law. (Typical error a few % for one quasar, but all the previous errors may make the whole distance scale in error by 20%.)
Learned: apparently universe exploded 13 to 20 million years ago.

LABORATORY EXERCISE VI

Properties of Pulsars and the Interstellar Medium

Kurtiss J. Gordon
Hampshire College

On a clear night, our eyes are dazzled by stars. A telescope shows us, as it did Galileo, that the hazy path we call the Milky Way "is, in fact, nothing but a congeries of innumerable stars grouped together in clusters." Sir William Huggins' spectroscopic observations in the 1860's showed, for the first time conclusively, that some of the hazy patches in the sky, or nebulae, were clouds of tenuous incandescent gas surrounding clusters of stars. Harlow Shapley argued, starting around 1916, that the model of the Milky Way derived from the method of "star gauges" was seriously in error. This method involved counting stars of various apparent brightnesses in different directions and assigning them distances on the basis of the inverse square law under the assumption that space is perfectly transparent. It had been pioneered in the 1870's by Sir William Herschel, and was used with increasing refinement as late as a 1922 study by J. C. Kapteyn. In spite of Shapley's arguments, it was not until the work of R. J. Trumpler in the 1930's that the existence of a generally distributed interstellar medium, which dims and reddens the light of distant stars, became widely accepted. Trumpler showed that, if one didn't adjust the calculated distances to stars for the effects of interstellar absorption, then the computed sizes of star clusters appeared to increase in all directions systematically with the distance of the cluster from the sun.

LABORATORY PART 1

The Interstellar Medium

Today we believe the interstellar medium contains dust (solid grains typically of the order of a micron in diameter), gas (molecules, atoms, ions, and free electrons), cosmic rays (electrons and atomic nuclei travelling at speeds close to the speed of light), magnetic fields, and, of course, photons (radio waves, light, x-rays, etc.). Almost everything we know about the interstellar medium--and about the stars--comes from collecting and analyzing the photons with our telescopes. Most of the mass of the interstellar material is bound up in the gas and dust, however; and dynamical studies of the Galaxy indicate that the interstellar material may comprise nearly as much of the Galaxy's mass as do the stars. Our data on the gas, dust, and magnetic fields in the interstellar medium consist of the ways they have interacted with the photons we collect. Since the photons are produced by stars and radio sources, we must understand something about the characteristics of the sources before we can say how the photons have been changed by their travels through the interstellar medium.

LABORATORY EXERCISE VI

Pulsars

Pulsars are objects, first detected in 1967 at Cambridge University's Mullard Observatory, from which we receive pulses of radio waves. The pulsar period, which is the length of time between successive pulses, is constant almost to the precision of our best atomic clocks. Figure 1 (see end of lab) contains recordings of the power received at the 140-foot telescope of the National Radio Astronomy Observatory in Green Bank, West Virginia, from three different pulsars. For each pulsar we show recordings made simultaneously at several different radio frequencies.*

---

*The radio frequency is the number of radio waves that reach the observer each second. A frequency of one wave per second is called 1 hertz; one of the frequencies used in the recordings in Figure 1 is 234 megahertz (MHz) or 234 million waves per second. For any kind of wave, one can also define a wavelength--the distance between corresponding points on neighboring waves. Frequency (f) and wavelength (λ) are inverse of each other. They are related by the speed at which the wave travels. Thus, for light or radio waves, travelling through a vacuum, $f(\lambda) = c$ (the speed of light). Radio waves with a frequency of 234 MHz have a wavelength of 1.28 m, slightly over four feet. Optical astronomers have traditionally thought in terms of wavelengths and radio astronomers in terms of frequencies. The reason for this is not profound, just a matter of which property of the wave is easier to measure in the two different regions of the spectrum.

* * * * *

(A) Determine the period of each of the pulsars--the length of time between successive points--by measuring the distances between pulses on the recordings in Figure 1. Try to estimate to tenths of a millimeter. You can achieve maximum accuracy by measuring the distance between widely separated pulses, and then dividing your result by the number of pulsar periods covered by your measurement. (Why is this true?) In determining the number of periods between the pulses you choose to measure, do not be misled by the fact that some of the intervening pulses may have been too weak to show up at all on the recordings. (This effect is particularly evident for PSR 0950+08.) In order to convert the period measurements from millimeters to seconds in time, measure the distance in millimeters between tick marks a known number of seconds apart. The tick marks are located at the top and bottom of the recordings of each pulsar. You should find a scale of about 50 mm/sec, but the exact value will depend upon what happened during the printing process. Divide the measured periods by the scale factor to obtain the periods in seconds. Note that the pulsar period is characteristic of the pulsar, and does not depend on the observing frequency.

* * * * *

To account for the pulsar phenomenon, theoreticians have proposed models based on neutron stars--the incredibly dense and compact remnants of the cores of stars that have undergone supernova explosions. These models use the rotation of the neutron star (sort of a cosmic "lighthouse," see Figure 2) to explain both how very precise the pulsar period is and why the period is the same for all observing frequencies. When a neutron star forms, a stellar core with a mass roughly equal to the mass of the entire sun collapses to become an object perhaps 10 miles in diameter. Like an ice skater pulling in his or her arms, the collapsing star spins more and more rapidly as it shrinks. The star's magnetic field is squeezed together, becoming very powerful--perhaps a billion times more intense than the field that holds the earth's Van Allen belts in place. The radio waves, which probably come from a small region on or above the surface of the neutron star, are beamed into a narrow cone by the powerful magnetic field. As indicated in the figure, the radio beam is swept around by the star's rotation, and what we detect as the pulses from the pulsar is the beam from this "lighthouse" sweeping past the earth.

Figure 2:

Schematic model of a pulsar.

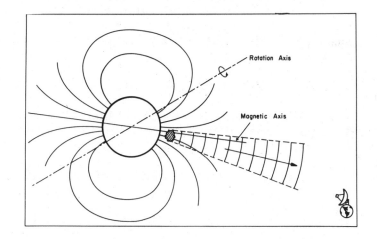

Pulse Dispersion and the Distance of Pulsars

The strength of successive pulses varies quite dramatically and rather erratically, although, for some pulsars, some order can be discerned amidst the chaos. For any pulsar, however, an individual pulse that is stronger than the average at any one frequency or wavelength will also be stronger than the average pulse from that pulsar at other frequencies over a very broad range of frequencies. This fact makes it possible to identify particular pulses observed with receivers tuned to different frequencies even though the pulses do not appear at the same instant in the recordings from the different receivers. The behavior just described can be seen in Figure 1. For PSR 0809+74, the arrival of one of the pulses is marked at the three different frequencies. Since time increases to the right in this diagram, we can see that the pulses from the pulsar arrive at the telescope earliest at the highest frequency, and progressively later at lower and lower frequencies. We say that the pulsar signals are dispersed.

LABORATORY EXERCISE VI

The dispersion of pulsar pulses gives one way in which they can be distinguished from terrestrially produced interference such as bursts of lightning, which arrive at the same time at all frequencies. The simultaneous arrival of a burst of interference at the three observing frequencies is also marked on the recording of PSR 0809+74. Detection of dispersed pulses is one of the methods that has been employed in searches for pulsars.

The dispersion of the pulsar pulses occurs as the radio waves travel through the interstellar medium from the pulsar to us. The radio waves interact (predominantly) with the electrons in the medium, and travel more slowly than they would through a perfect vacuum. Lower frequency waves are affected more strongly by a given number of electrons, so they fall behind the higher frequency waves.* For any pair of frequencies, the observed delay--the actual length of time (Δt) between the arrival of the pulse at the higher frequency and the arrival of the same pulse at the lower frequency--is proportional to the number of electrons in between the pulsar and the observer. The number of electrons is just the product of the average concentration of electrons (n_e) in each cubic centimeter of space and the distance (d) to the pulsar. The quantity $n_e d$ is often called the <u>dispersion measure</u> (DM) of the pulsar. The actual equation governing the delay between frequencies f_1 and f_2 is

$$\Delta t = 4150 \, n_e d \, (1/f_1{}^2 - 1/f_2{}^2), \qquad (1)$$

where the constant is chosen so that Δt is measured in seconds, n_e in electrons/cm^3, d in parsecs (1 pc = 3.26 light years), and f_1 and f_2 in megahertz.

In general, more distant pulsars should send us more highly dispersed pulses. If we know n_e from other information, we can measure Δt and solve for the distance to the pulsar. Even if we don't know the actual value of n_e toward a particular pulsar, we can assume a typical value (usually $n_e = 0.03$ cm^{-3}), and get a rough estimate for d. But, over distances of a few hundred parsecs, these variations appear to average out, so that the distance estimates for most pulsars are probably good to within a factor of approximately 2.

_____
*The phenomenon of dispersion is the same one that permits a glass prism to separate light of different colors into a spectrum. However, in glass the higher frequency light waves (violet colors) travel more slowly than the lower frequencies (red colors), while in the interstellar medium <u>the lower frequency</u> radio waves travel more slowly.

* * * * *

(B)   Determine the dispersion delay, Δt (the length of time by which the pulse at a lower frequency is delayed), for PSR 0329+54 and 0950+08 at several pairs of frequencies. Using either a ruler, or a translucent precision two-millimeter grid laid over the recordings, measure the distance from a reference line to the peaks of the corresponding pulses at each of the observing frequencies. If you are unsure about which pulses correspond to one another, consider the pattern of relative strengths of successive pulses at each frequency. Try to estimate to tenths of a millimeter. To calculate Δt in millimeters, take differences between the measurements for a particular pulse at any two different frequencies. For higher accuracy, use more than one set of corresponding pulses and average your results at the same pair of frequencies. (There are three different frequency pairs possible for PSR 0950+08, and six possible pairs for PSR 0329+54.) Convert Δt from millimeters to seconds of time by using the scale factor you found in exercise (A).

* * * * *

(C)   Determine approximate distances to PSR 0329+54 and 0950+08. Use equation (1) to calculate the distances from the dispersion delays (Δt) measured in exercise (B). You may assume that n_e = 0.03 electrons/cm^3. To save you some arithmetical manipulations, here is a list of values of the quantity $4150 \, (1/f_1{}^2 - 1/f_2{}^2)$ which appears in equation (1):

f_1	f_2	$4150(1/f_1{}^2 - 1/f_2{}^2)$	f_1	f_2	$4150(1/f_1{}^2 - 1/f_2{}^2)$
[MHz]	[MHz]	[cm^{3}pc^{-1}sec]	[MHz]	[MHZ]	[cm^{3}pc^{-1}sec]
234	256	1.247×10^{-2}	256	405	3.802×10^{-2}
234	405	5.049×10^{-2}	256	1420	6.127×10^{-2}
234	1420	7.373×10^{-2}	405	1420	2.324×10^{-2}

The slightly different distances you obtain from the three pairs of frequencies for the same pulsar represent the difficulty in measuring Δt exactly. (Should any pair of frequencies give higher accuracy than any other pair?)

* * * * *

Although we used an average value of n_e and determined the distances to the pulsars, the procedure may be turned around. If we have some other way of knowing the distance to a pulsar, we can solve equation (1) for n_e in this particular direction in space. For example, the pulsar PSR 0531+21, with DM = 56 cm^{-3}pc, has been identified as the central star of the Crab Nebula. Analysis of the expansion of the nebula leads to a distance estimate of d = 2000 pc. Since the DM = $n_e d$, the electron concentration n_e = 0.028 cm^{-3}, on the average, toward this pulsar.

LABORATORY EXERCISE VI

LABORATORY PART II
(for advanced students)

In the conceptual sequence of this laboratory, the following two exercises fall most logically before exercise (B). Exercise (D) is the most difficult of the group, and should be assigned to the most advanced students. Either or both of these exercises may be omitted for other students without disturbing the continuity of the laboratory.

* * * * *

(D)   Derive equation (1). The velocity of propagation, or <u>group velocity</u> (u), of radio waves through an ionized gas like the interstellar medium is given by

$$u = c \sqrt{1 - \text{const} (n_e/f^2)}, \qquad (2)$$

where c is the velocity of light and radio waves in a vacuum and

$$\text{const} = e^2/(4\pi^2 \varepsilon_o m) = 8.06 \times 10^7 \text{ cm}^3/\text{sec}^2.$$

Using the familiar relation that distance traveled = velocity $\times$ time, with velocities given by equation (2), compute an expression for the dispersion delay, $\Delta t = t_1 - t_2$ (the difference in travel time for waves of two different frequencies f_1 and f_2). To simplify the resulting equation, expand it in a power series according to the binomial theorem and ignore all terms after the second.

* * * * *

(E)   Satisfy yourself that equation (1) works for a patchy medium by dividing the interstellar medium along the line of sight to the pulsar into several regions with different n_e and summing the delays accumulated within each of the regions.

* * * * *

The following exercise may be inserted between exercise (B) and (C).

(F)   If equation (1) holds true for all frequencies, and if pulsars were detectable optically, would you be able to measure dispersion for optical pulses on recordings similar to the ones in Figure 1? The frequencies of violet and red light are 8×10^8 MHz and 4×10^8 MHz, respectively.

* * * * *

Polarized Radio Waves and the Phenomenon of Faraday Rotation

Because the vibrating electric and magnetic fields that make up a radio wave are directed transversely to the line of sight, it is possible to define a plane of polarization for the radio wave. To understand this let us consider an analogous case of a wave traveling along a length of rope. If you and a friend hold opposite ends of a rope fairly tautly, and you shake your hand rapidly up and down, a vertically polarized wave will travel down the rope to your friend. If you shake your hand from side to side, you will send your friend a horizontally polarized wave. With no obstructions between the two of you, you are free to send your friend a wave of any desired polarization. However, if the rope passes between the slats of a picket fence, only vertically polarized waves or the vertical component of obliquely polarized waves will reach your friend. The picket fence acts as a polarizing screen, which is transparent to waves of only one polarization. The same principle is used in Polaroid sunglasses. Since sunlight contains roughly equal numbers of light waves of all possible polarizations, about half of the waves are filtered out and only the components of light waves with a polarization appropriate to the alignment of the polaroid are transmitted. Unlike the sun, many pulsars emit radio waves which are predominantly of one polarization to begin with.

When radio waves of a particular polarization traverse an ionized medium in which there is a magnetic field, the plane of polarization rotates around the line of sight. This phenomenon is called Faraday rotation after its discoverer, the 19th century English physicist Michael Faraday. Thus, radio waves emitted by a pulsar with their plane of polarization oriented N-S may, for example, arrive at the observer with their plane of polarization oriented NE-SW. The angle ($\Delta\Theta$) through which the plane of polarization is rotated depends upon the frequency of the radio waves (f), the electron concentration (n_e) in the interstellar medium, and the distance travelled (d). It also depends on the strength of the component of the interstellar magnetic field along the line of sight ($B_\ell = \cos\phi$ for a magnetic field of strength B oriented at an angle ϕ with respect to the line of sight). Specifically,

$$\Delta\Theta = 4.18 \times 10^{12} \, n_e \, B_\ell \, d/f^2 . \qquad (3)$$

The value of the constant has been chosen to allow us to measure $\Delta\Theta$ in degrees, B_ℓ in gauss, and n_e, d, and f as before in electrons/ cm^3, parsecs, and megahertz, respectively. Since the B_ℓ component of the magnetic field may be directed towards or away from the observer, the plane of polarization of the radio wave may be rotated in either a counterclockwise or clockwise direction. The traditional measurement of angles on the sky is from N through E (counterclockwise). A counterclockwise Faraday rotation is produced by a magnetic field directed towards the observer. Consequently, clockwise rotation and fields directed away from the observer are traditionally defined as negative.

If a radio source emits waves of all frequencies, all polarized in the same direction, then it is theoretically possible to determine the intrinsic plane of polarization (Θ_o) simply by observing at an infinitely high frequency. According to equation (3), waves of infinite frequency should not be Faraday rotated at all. In practice, we observe at finite frequencies, so we do not have an <u>a priori</u> knowledge of Θ_o. If Θ_o is in fact the same at all frequencies, then a plot of the observed plane of polarization ($\Theta = \Theta_o + \Delta\Theta$) against $1/f^2$ will be a straight line. This line can be extended, or extrapolated, to $1/f^2 = 0$ in order to find Θ_o. Note, however, that an N-S orientation of the plane of polarization can be described as $\Theta = 0°$ or 180° or 540°, etc. Therefore, it is legitimate, and may be necessary, to add or subtract some multiple of 180° to any of the observed values of Θ in order to achieve a straight line on the graph.*

_____

*Harking back to the rope and picket-fence analogy may help to show why the ambiguity occurs every 180°, and not after a full 360°. You can produce vertical waves on a rope equally well by shaking your hand up-and-down or down-and-up. Either way, the waves can penetrate the picket fence. Indeed, except for the initial half cycle of the wave, there is no difference between the two. In the astronomical case, then, there is no observable difference between N-S and S-N polarized radio waves.

Because of this ambiguity, observations in at least three different radio frequencies are needed to determine the amount of Faraday rotation that radio waves from a particular radio source have undergone.

* * * * *

(G) Table 1 contains observed values of the orientation of the plane of polarization of radio waves of several different frequencies from some pulsars. For each pulsar, determine Θ_o and $\Delta\Theta$ (at 400 MHz). You will find it easiest to start with the data for PSR 0950+08. For PSR 0329+54, work outward from the two closest-spaced frequencies (410 and 414 MHz), because the full range of the data spans several complete rotations of the plane of polarization.

Because of the observational difficulties in measuring Θ, the values tabulated for each pulsar will not lie precisely on a line of the form of equation (3). However, none of the points should deviate more than about ± 10° from the best-fitting line. (How do the uncertainties in the values of Θ affect the accuracy with which you can determine Θ_o?)

Table 1

Observations of Pulsar Polarization

f	$1/f^2$	Observed polarization angle (θ) for PSR		
[MHz]	[10^{-6}MHz^{-2}]	0239+54	0809+74	0905+08
280	12.76	–	–	97
281	12.66	–	30	–
365	7.51	130	0	54
392	6.51	103	54	55
410	5.95	100	94	36
414	5.83	129	–	–
421	5.64	26	103	38
485	4.25	114	–	–
1665	0.36	155	–	5

* * * * *

Rotation Measure and Interstellar Magnetic Fields

Multiplying both sides of equation (3) by f^2 allows us to define a rotation measure (RM), which is independent of frequency and depends only on factors describing the conditions in the interstellar medium between the pulsar and the observer.

$$RM = 2.39 \times 10^{-13}f^2 \, \Delta\theta = n_e B_\ell d = B\ell DM, \qquad (4)$$

with RM defined in these units, the ratio of two directly observable quantities (RM/DM) is a measurement of the strength of the interstellar magnetic field.* Actually, it yields a lower limit to the field strength for two reasons. First, Faraday rotation depends only on the component of the magnetic field along the line of sight. Second, if B_ℓ is toward the observer along part of the line of sight and away from the observer along the rest of it, the Faraday rotation will be in opposite directions in the two regions. The effects of the two regions will partially offset each other, and the observed rotation will be smaller in magnitude than the rotation produced in either region separately. The advantage of observing Faraday rotation of pulsars rather than other radio sources is that only for pulsars is DM an observable quantity. Hence, only for pulsars can one obtain an estimate of B_ℓ without making assumptions about the values of n_e and d.

---

*Rotation measures are traditionally defined in units of radians/m^2 from the equivalent equation $RM_{trad} = (\theta_1 - \theta_2) / (\lambda_1^2 - \lambda_2^2)$, for the observing wavelengths λ_1 and λ_2. Because of the different units, $RM = 1.23 \times 10^{-6} \, RM_{trad}$.

* * * * *

(H)   Calculate RM's for the pulsars analyzed in exercise (G).   Use the DM's from exercise (B) to turn these into estimates of B_ℓ.

* * * * *

Observations of Faraday rotation of the radiation from pulsars and from extragalactic radio sources in all directions have been used to map out the large-scale structure of the galactic magnetic field. Within a region around the sun, several thousands of light years in diameter, the fields appear to have a roughly constant strength of a few microgauss (millionths of a gauss; about 10^5 times weaker than the magnetic field near the surface of the earth).   Its form is longitudinal, according to these observations, and it is directed nearly parallel to the direction of the sun's revolution around the center of the Galaxy--or perhaps, along the spiral arms which deviate by only $6°$-$12°$ from that direction.   Superimposed on the longitudinal field are a few irregularities, also with strengths of a few microgauss. The most noticeable of these is associated with the Galactic Spur, a major feature in the galactic radio continuum emission.

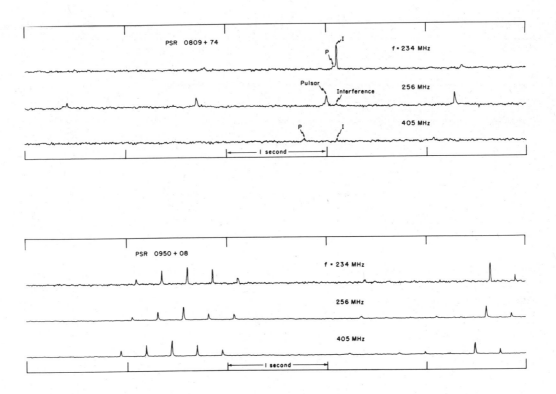

Figure 1(a, b)   Observations of the pulsars PSR 0809+74 and 0950+08 at three radio frequencies show the radio power received from the pulsars as a function of time.   Tick marks at intervals of one second appear above and below the recordings for each pulsar.   In the recordings, time increases from left to right, and the power received increases upward.

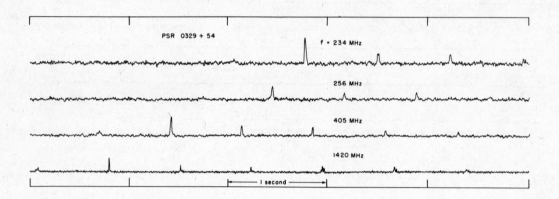

Figure 1(c)   Observations of the pulsar PSR 0329+54 at four radio frequencies.   See the caption to parts (a, b) of this figure for more description.

LABORATORY EXERCISE VI

Note for Teachers: The three exercises in Part I of this lab should form an assignment that can be completed by introductory students in a single two-hour laboratory session, allowing some time for introductory remarks.   For more advanced students, you can add exercises from Part II, perhaps omitting exercise (A).   You might also ask the students to estimate the uncertainty of each of their measurements and carry these uncertainties through their calculations to see how much effect they could have on the answers.   Students with an advanced e & m background might be asked to derive equations (2) and/or (3).   Completion of all exercises in both parts of the lab will probably require two two-hour sessions, even without the additional suggestions in this note.   By appropriate choice of the material, you can construct a lab that is interesting and challenging to students at any level from freshman to first-year graduate school.

Answers:

Published values for the periods and dispersion measures are:

PSR:	0329+54	0809+74	0950+08
P[sec]	0.71452	1.29224	0.253065
DM[$cm^{-3}pc$]	26.776	5.757	2.969

The following dispersion delays and distances are computed from the published DM's.   For PSR 0329+54, notice that the Δt for the adjacent frequencies 256 and 405 MHz is longer than the pulsar period.

$\Delta t(f_1 f_2)$[sec]	0329+54	0809+74	0950+08
234, 256	0.333	0.072	0.037
234, 405	1.351	0.291	0.150
234,1420	1.973		
256, 405	1.018	0.219	0.113
256,1420	1.640		
405,1420	0.622		
d [pc]	893	192	99

A least-squares solution to the data in Table 1 after visual removal of the ambiguity gives:

Θ [°]	90	102	-1
$\Delta\Theta_{400}$[°]	-2086	-390	47

Acknowledgments:
     I would like to thank Dr. R. N. Manchester for supplying the polarization data summarized in Table 1 from his observations at the NRAO 300-foot telescope.   I am very grateful to Drs. C. P. Gordon, G. R. Huguenin, and R. E. White, and to the students in our classes, for trying out and commenting on the preliminary versions of this exercise.   The NRAO is operated by Associated Universities, Inc., under contract with the National Science Foundation.   This exercise was developed with partial support from a VITAL (Ventures in Teaching and Learning) grant from the IBM Corporation to Hampshire College.

LABORATORY EXERCISE VII

Observing Lab

Note: Each person, of course, will want to use an observing program that makes the best use of the school's observational facilities. The following is the program that we use at Williams College.

At the beginning of each semester, each student receives the following handout:

Observing Program

The purpose of this program is to help you to become familiar with the types of objects that can be found in the sky and to introduce you to some of the techniques that astronomers use to observe these objects and gather the information contained in the their light. You will have the opportunity to do both daytime and nighttime observing and although we have stated below the minimum number of visits you must make, you are encouraged to come for observing as often as you like. If you have any questions about the program, please do not hesitate to ask.

Daytime Observing

Each student will observe the sun on at least two occasions, separated by a period of between two and five days (weather permitting). On each occasion you will view the sun in two different types of light and note the differences between the two solar images. You will also see the changes in the face of the sun over the time°° between your visits. You are requested to make careful pencil sketches of all the observations you make on the sun; paper is provided if you need it. ("Sketches" are just that--they are not finished works of art and they are meant to be rough. We are not asking you to be artists; we just want you to learn to record what you see.) There will be two telescopes set up for solar observing: a Celestron 8 equipped with a special kind of filter that allows one to safely view the white-light solar image, and a five-inch refracting solar telescope with an H-alpha filter mounted on it, for viewing the sun in the light of one particular wavelength produced by hydrogen (called H-alpha). The solar spectrum will also be on view.

Daytime observing hours are from 1:30-2:30 p.m., Monday through Friday. Students are free to choose the times of their visits, but are encouraged to complete all their observing as early as possible in the semester, since the weather is seldom cooperative in November. If you find these times inconvenient, additional times can be arranged, but please tell us about this as early as possible.

LABORATORY EXERCISE VII

Nighttime Work

Each person will pay at least TWO nighttime visits to the telescopes. ONE visit will be for photography (see below) and may be made either when the moon is visible and in good position for photographing, or when the moon is not up--keep in mind that the presence of the moon makes other normally observable objects difficult to see because its brightness past first quarter tends to wash out the night sky. The other visit will be for observing and taking notes and will be made during the opposite part of the lunar cycle. We will keep you informed of the moon's phases, and you are encouraged to keep track yourself. Types of objects that can be observed (primarily when the sky is dark) include planets, star clusters and multiple star systems, nebulae, galaxies, and whatever else the night assistants can pick up.

Equipment for nighttime observing includes two Celestron 8 reflectors (8" = 20-cm diameter telescopes of the Schmidt-Cassegrain type), the Celestron 14 reflector (a 35-cm Schmidt-Cassegrain), a 20-cm Celestron Schmidt (that is, with a wide field), and the "solar telescope," which can be converted to nighttime use. In addition, there are binoculars available for students, who are welcome to bring their own if they wish.

The hours for nighttime observing are from 8:30-9:30 p.m., Sunday through Friday at the beginning of the semester; later in the semester it gets dark earlier and new hours will be announced. Again, students are free to do their observing when they wish, provided it is clear. (Observing clouds is not a very interesting way to spend an evening.) Sketches of nighttime observations are encouraged but not required. Each student must keep a notebook listing (1) the dates of his or her observations, (2) the objects observed, (3) the quality of the sky (haze, partial clouds, wind, temperature, etc.), and (4) what you saw. A log is kept on the roof for students to sign whenever they visit, to keep track of who has and has not observed.

Photography

Each student will take a sky photograph of an object of his or her choice (of course, there are limits to what our equipment is capable of--you will be guided in your selection).

We do three different types of sky photography; each type involves a different photographic set-up. When the moon is in a good position to be photographed, we use a 35mm camera mounted at the focus of the Celestron 8. Because of the moon's brightness we use an extremely slow film, and exposures are timed with a hand-held shutter. The method has proven quite effective in obtaining excellent lunar photographs.

Planetary photography is a somewhat trickier business. The photographic set-up is only slightly different from that used in lunar photography; the 35mm camera back with slow film is used in conjunction with an eyepiece (usually 25 or 12.5mm, depending on sky conditions) at the focus of the Celestron 14, and the exposure is again

LABORATORY EXERCISE VII

timed by hand. It is difficult to get good results in planetary photography. No planets are up in the evening early in the semester.

Deep-sky (dark of the moon) photography has turned out to be a particularly rewarding activity for us; we use an 8-inch Schmidt camera mounted on the Celestron 14 with a single frame of 35mm film at the Schmidt's prime focus and the Celestron 14 itself. The Schmidt is a specially built camera, designed to give sharply defined images over a wide field of view (4.5 × 6.5 degrees--the moon subtends .5 degree). The photographer guides the camera during the entire exposure by viewing through the Celestron 14 and making small corrections in the camera's tracking using the telescope's fine motion controls. Deep-sky work tends to be lengthier and more involved than either of the other types, but if gone about carefully, the results are usually very good.

Scheduling for Photography

Because of the number of people in the course, and because we anticipate that a hefty portion of the nights will be unusable, we must schedule you close together and can afford only 30 minutes for each photograph. There will be five students scheduled for photography each night, with sessions beginning at 9:00 and going until about 11:30 p.m. Students may choose when they want to photograph; our only condition is that if you choose to do your photography when the moon is present, then you choose your other observing period when the moon is absent, and vice versa. This will ensure that everyone gets up to the roof once during the light and once during the dark of the moon. Students must sign up in advance for the time they want to photograph; signing up will take place during the week for the following week's sessions.

After the first several weeks of the semester, everyone will have had a chance to sign up for photography. Naturally, not everyone will actually have gone through the process, as many of the nights will have to be cancelled. Those students who have not photographed will continue to sign up until they hit a clear evening. You may sign up as often as you need to until you have completed the photography. In this way we hope to enable each student to obtain his own photograph, and to give each student an idea of what astro-photography involves. In order to carry out this program successfully, we need to have the cooperation of all of you. PLEASE DO NOT FAIL TO APPEAR AT THE TIME FOR WHICH YOU HAVE SIGNED UP and please have patience if you keep landing cloudy evenings--even the most efficiently designed observing program is necessarily limited in its effectiveness by the weather, and that is especially true in the Berkshires.

Photography sessions will be held six nights a week, Sunday through Friday. You will be informed of any changes in the program, if they occur. Again, if you have any questions about anything, please ask. If you have difficulty with your scheduling, please inform us as soon as the problem arises, and we will try to work things out with you.

LABORATORY EXERCISE VII

_____ is in charge of the observing deck. S/he will be assisted by one or two student night assistants. They are there to help with observing and especially with photography, and they will point out things, explain things, and answers questions for you. Their job is specifically to help you, so please go to them first with your problems.

Good luck, and have fun! Note that if you want to learn the constellations just ask about them on your observing visits. Constellation study is optional, not mandatory, but you may have as much of it as you like.

Lab notebooks will be collected in class on December 1. Students must have completed both the 2 daytime and the 2 nighttime visits to the observing deck in order to pass the course. Note that it is not unheard of to have three straight weeks of clouds in November, so do try to come to observe in September or October.

LABORATORY EXERCISE VII

Name _____

Observing Checklist

Fill in the date for each object observed.

I. The Sun   1st visit: white light _____   Hα _____
                              spectrograph _____
                  2nd visit: white light _____   Hα _____
                              spectrograph _____

II. Light Time   Moon _____   phase: _____

                  Double star _____   Name _____
                  Planet _____   Name _____
                                      Name _____
                  Constellations _____   Name _____
                                      Name _____
                                      Name _____
                                      Name _____
                                      Name _____

III. Dark Time   Double star
                     Albireo _____
                     Eta Cassiopeia (η Cas) _____
                     Mizar _____
                     Gamma Cygni (γ Cyg) _____
                     Other _____
                  Globular Cluster
                     M13 in Hercules _____
                     M15 in Pegasus _____
                     Other _____
                  Galactic (Open) Cluster
                     h and chi (χ) Persei _____
                     Pleiades _____
                     Hyades _____
                     Praesepe (Beehive); M44 _____
                     M35 in Gemini _____
                     Other _____

                  Planetary Nebulae
                     Ring _____
                     Dumbbell _____
                     Other _____
                  Other Nebulae
                     Orion (winter sky) _____
                     North America _____
                     Other _____

LABORATORY EXERCISE VII

Galaxy
    Andromeda _____
    Whirlpool (M51) _____
    Other _____
Milky Way
    Northern Coal Sack _____

Supernova Remnant
    Crab _____
    Cygnus Loop _____
      (including Veil)

IV. Meteor   _____
                   _____

V. Bright Stars
    Sirius   _____
    Arcturus   _____
    Vega   _____
    Capella   _____
    Rigel   _____
    Betelgeuse _____
    Procyon   _____
    Aldebaran   _____
    Polaris   _____

VI. Variable stars   (observe at two dates)
    Delta Cephei _____ and _____
    Mira   _____ and _____
    Algol   _____ and _____

This summary sheet should be turned in along with your lab notebook.

Note: Spelling counts on this sheet and in your lab book. Check correct spellings of bright stars, Messier Objects, and constellations in the appendices.

LABORATORY EXERCISE VIII

Magnitudes

Students often have trouble realizing just what it means for magnitudes to be on a logarithmic scale, especially now that calculators have replaced the use of logarithms. James A. Blackburn of the Physics Department, Wilfred Laurier University, Waterloo N2L 3G1, Ontario, Canada, has designed an electronic circuit that allows students (or the lecturer) to adjust six lights so that they show equal intervals of brightness as seen by the eye. Oscilloscope readouts, however, show that the intensity is logarithmic.

Prof. Blackburn described his device in an article entitled "Integrated Circuit Stellar Magnitude Simulator," which appeared in the American Journal of Physics 46(8), August 1978, pp. 813-814. We have built one at Williams following his design, and used it for a classroom demonstration. It could also be used for a student laboratory. The cost of the parts is in the $35.00 range.

LABORATORY EXERCISE IX

The Analemma

Robert Allen
Physics Department
University of Wisconsin, La Crosse, WI 54601

The analemma is a graphical representation of the position of the sun during the course of the year. On many globes of the world, you may find a figure-8 shape, usually located in the mid-Pacific Ocean where there are few geographical objects; this is the analemma.

General Information

Notice that along the left side of the analemma graph are numbers indicating the declination of the apparent sun. The declination of the apparent sun ranges from 23.5° north to 23.5° south (or -23.5°) of the celestial equator. The exact value of the declination for any date can be read from the analemma. For example, on October 14, the declination of the apparent sun is given as 8° south, while on August 10, the declination is 16° north.

Along the top of the graph of the analemma are given the values of the "Equation of Time," which tells the amount of time before or after noon (local mean time) that the apparent sun will cross the upper meridian. The values range from 16.5^m "Sun fast" to about 13.25^m "Sun slow." The basic purpose of this laboratory is to learn to read the analemma.

Declination

Declination is the angular distance of an astronomical object north or south of the celestial equator. For a person at the equator of the earth, the celestial equator passes exactly overhead through the zenith; consequently, for a person at the earth's equator, the altitude where the upper meridian intersects the celestial equator is _____ . For a person at the earth's North Pole, the altitude where the celestial equator intersects the upper medidian is _____ . Our latitude is _____ North. Consequently, the altitude of the point where the upper meridian intersects the celestial equator for a person here is _____ . Suppose for a given date the declination of the apparent sun is 10° N; then for a person here, the altitude of the sun when it reaches the upper meridian would be _____ .

For the dates given below, obtain from the analemma the declination of the apparent sun, and also determine the altitude of the apparent sun (when on the upper meridian) for an observer at our location and at the earth's equator.

LABORATORY EXERCISE IX

DATE	DECLINATION	NOON ALTITUDE (for us)	NOON ALTITUDE (at equator)
January 12			
April 11			
June 21			
October 21			

General Questions on Declination and Altitude

1. Determine from the analemma two dates on which the declination of the apparent sun is -16° (that is, 16°S):

----------------------------   ----------------------------

2. Determine two dates on which the altitude of the apparent sun when on the upper meridian as measured here is 38°:

----------------------------   ----------------------------

Equation of Time

The equation of time provides the information which changes local apparent solar time to local mean solar time. In order to understand the analemma's information in detail, it is necessary to get a few definitions straight.

Mean sun—This is a fictitious point which moves at a constant rate along the celestial equator, completing exactly one circuit each year.

Local solar mean time—This time is determined in exactly the same way as sidereal time, except that instead of using the vernal equinox it uses the "mean sun." Consequently, local mean solar time is the amount of time since the "mean sun" crossed the meridian.

Local apparent solar time—This time is the same as the local mean solar time, except that it uses the real or apparent sun rather than the mean sun. Consequently, local apparent solar time is the amount of time since the real or apparent sun last crossed the meridian.

LABORATORY EXERCISE IX

We can now state the meaning of the equation of time. Using ET as the symbol for the equation of time, LAT as the symbol for local apparent time, and LMT as the symbol for mean local time, we can write the following equations:

$$ET = (LAT) - (LMT) = (AT) - (MT)$$

or

$$LMT = (LAT) - (ET).$$

The reason this equation becomes necessary is because the real or apparent sun does not require exactly the same amount of time every day of the year to proceed all around from the upper meridian to setting to the lower meridian to rising and back to the upper meridian. Some days of the year, a solar day is as much as 20 seconds more than 24 hours, and some days of the year, it is as much as 20 seconds less than 24 hours. Consequently, local apparent solar time (which uses the real sun) would make a very poor clock, because you would have to continually correct your watch in order to maintain accurate time.

Local mean solar time averages out the various long days and short days during the year. The information on the analemma provides the means by which you can determine whether the apparent sun will cross the upper meridian before (sun fast) or after (sun slow) 12:00 noon local mean solar time, and by how much. Check your analemma to see that on October 21, the apparent sun will cross the upper meridian 15 minutes before noon local mean solar time; and that on March 29, the apparent sun will cross the upper meridian 5 minutes after noon, local mean solar time. For the dates given below, determine whether the apparent sun will cross the upper meridian before or after noon, local mean solar time, and by how much.

DATE	BEFORE OR AFTER NOON LOCAL MEAN SOLAR TIME?	AMOUNT OF TIME BEFORE OR AFTER?
January 21		
April 11		

Correction from Local Time to Standard Time

In these notes on the equation of time to this point, there has been an attempt to emphasize the fact that the time corrections obtained from the analemma are corrections to what has been called local mean solar time. If you wish to determine the actual time at which the apparent sun will cross the upper meridian for an observer at a particular place on earth, you must make the additional correction from local time to standard time. You should _____ (add or subtract) _____ minutes in order to change a (your location) local mean time determination to a (your time zone) Standard Time. Since the analemma provides only corrections to local mean solar time, to determine __ST you must make this correction.

LABORATORY EXERCISE IX

For the following dates, determine the _____ Standard Time at which the apparent sun will cross the upper meridian here.

DATE	EQUATION OF TIME VALUE	LOCAL APPARENT TIME	LOCAL MEAN TIME	__ST
August 12				
November 28				
June 3				

General Question on Time

Find four dates on which the apparent sun will cross the upper meridian at 12:07 p.m. __ST for an observer here.

_____   _____   _____   _____

General Questions on Time and Declination

The altitude of the apparent sun at _____ local apparent noon and the __ST at which the apparent sun crosses the meridian are given for dates below. What are the dates?

ALTITUDE	__ST	DATE
41°	12:16 p.m.	
29°	11:49 a.m.	

Determine the altitude of the apparent sun at local apparent noon, and the time (__ST) at which the apparent sun will cross the upper meridian for an observer here for the following dates.

DATE	ALTITUDE	APPARENT SUN TRANSIT TIME (__ST)
June 3		
February 12		
December 25		

LABORATORY EXERCISE IX

+ Apparent Sun Fast   Apparent Sun Slow —

Equation of Time

The analemma

LABORATORY EXERCISE X

The Graphic Time Table of the Heavens

Robert Allen
Physics Department
University of Wisconsin, La Crosse WI 54601

Introduction

The Graphic Time Table is published by Scientia, Inc., 1815 Landrake Road, Baltimore, MD   21204, (301) 828 5494. Simplified versions for the next   years are available in Menzel and Pasachoff, Field Guide to the Stars and Planets, 2nd ed., 1983. Graphs that display similar data are published yearly in Sky & Telescope.

The table is designed to read directly the local times of astronomical events for an observer of 40° north latitude. (Directions for finding local times of events at other latitudes are included in the accompanying handout.)   Since the times of most events at our location (latitude of _____ North) will be within a few minutes of the times of 40° north latitude, we will use the direct readings for 40°. However, a correction must be applied to convert the local times to standard times. This correction is the same amount of time for each event.

Local vs. Standard Time

When you read an accurate watch or clock, you are reading _____ Standard Time (or _____ Daylight Saving Time between the last Sunday in April and the last Sunday in October).   The continental United States is broken into 4 standard time zones.   These are the Eastern, Central, Mountain, and Pacific zones.   The Central Zone is centered on the 90° W (6^h west) longitude line. Pacific Standard Time is centered on the 120° W (8^h west) meridian. Eastern Standard Time is therefore centered on the _____ meridian, while Mountain Standard Time is measured along the _____ meridian.

If an observer is located exactly on the standard meridian of his time zone, standard times of events can be read directly from the table without correction. For most observers, though, a correction must be made from local to standard times.

We are in the _____ Time Zone. Our longitude to the nearest 1/2° is _____ W (i.e., we are _____° west of the Greenwich meridian). In order to determine the __ST of an event (or __DST), we must ask what the time is at that same instant for an observer exactly on the 90, 75 . . . west meridian. This imaginary observer (at ___ W. longitude) is _____ (east or west) of us in _____ by _____ (how many degrees)? Since local time is _____ (earlier or later) for an observer east of us (at that same instant) and _____ (earlier or later) for someone west of us, we must _____ (add or subtract) a _____ minute correction (remember that $1 = 4^m$) to our local time (as read from the table) to obtain the __ST of an event. When Daylight Saving Time is in effect, an additional correction of +1 hour must be applied to the standard time. (Remember that we "spring our watches ahead" in April and "let them fall back" in the fall.)

Mean Solar Time vs. Apparent Solar Time

Solar time measurements are governed by the sun. However, the sun, unlike our watches, does not move at an equal rate throughout the day or year. At some times of the year the sun is "ahead of" our watches and at others it is "behind." The difference in mean solar time (as read by our watches) and apparent solar time (as would be read by a sundial) may amount to as much as about 17 minutes. This means that the sun reaches the upper meridian up to as much as 17 minutes ahead of or behind noon as read by our watches (and it therefore also reaches the lower meridian by as much as 17 minutes ahead or behind midnight as read by our watches). To see how far ahead of or behind noon or midnight the sun reaches the upper or lower meridian, examine the wavy line which "snakes around" the vertical midnight line on the table. When the wavy line is to the right of the midnight line, the sun is behind your watch. When the wavy line is to the left of the midnight line, the sun is ahead of your watch. By reproducing the scale in the lower left corner of the table (or using a ruler) you can determine exactly how far ahead of or behind your watch the sun is on a particular day.

Uses of the Graphic Time Table

The GTT can be used to determine the rising and setting times of the sun, moon and naked-eye planets throughout the year. It also gives upper meridian crossing times (upper transit times) of the brighter stars and gives some interesting binoculars and telescope objects.

Note: Each horizontal line represents a Thursday night-Friday morning. To find times of events on other nights, lay a straight edge along the appropriate horizontal line. The dashes on the sunset and sunrise curves indicate one-day intervals. Reproduce the scale in the lower left-hand corner when determining the times of particular events, and try to estimate times to within 2 or 3 minutes. Also, indicate whether times are a.m. or p.m.

LABORATORY EXERCISE X

Events during the evening of Tuesday, April 8, and morning of Wednesday, April 9:

EVENT	LOCAL TIME	__ST
	(directly from table)	

Sunset

Evening twilight ends
(sun 18° below horizon)

Mars reaches upper transit

Venus sets

Regulus reaches upper transit

What is the sidereal time at
   midnight on this date?

Sun at lower transit
(crossing lower meridian)

Jupiter sets

Saturn sets

Morning dawn begins
(sun 18° below horizon)

Mercury rises

Sunrise

Venus rises

Moonrise and Moonset

The purpose of this part of the lab is to determine the time of moonrise or moonset and use the daily "rises" or "sets" circles. The small darkened circles represent moonset and the small clear circles are for moonrise. New moons are marked by large black circles and full moons by large clear circles.

Give moonset times for the following:

DATE	MOONSET (from GTT)	__ST

April 15-16 (evening of the
15th and morning of the 16th)

April 16-17

April 17-18

LABORATORY EXERCISE X

Give moonrise times for the following:

DATE	MOONRISE (from GTT)	__ST	__DST
May 1-2			
May 2-3			
May 3-4			

The Moon's Cycle of Visibility

What is the date of the next new moon? _____

What is the date of the full moon in May? _____

On the date of the full moon in May the moon sets at about _____ (sunset, midnight, or sunrise, or noon).

On the date of the full moon, the moon rises _____ (amount of time) _____ (before or after) sunset.

Each night thereafter it sets about 1 hour _____ (earlier or later).

In about a week after the full phase the moon is at the _____ phase and gets about _____ (sunset, midnight, sunrise or noon).

General Questions

On what day of 198_ is the earliest sunrise? _____

On what day of 198_ is the latest sunrise? _____

What is the longest day of 198_? _____

On what day of 198_ is the earliest sunset? _____

What is the shortest day of 198_? _____

On what date is the vernal equinox (also called the "First point of Aries") on the upper meridian (at upper transit) at midnight? _____

Other Questions

On what date does Venus have its "Greatest Brilliancy," _____, and is seen as an "evening" object? _____

On what dates does Mercury achieve its greatest elongations, when it is seen as a "morning" object? _____

Atmospheric Extinction

Roy H. Garstang
Department of Astro-Geophysics
University of Colorado, Boulder, CO   80309

For: Clear evening, not too hazy.

Purpose: To estimate very roughly how much light is lost as a result
          of absorption and scattering in the atmosphere.

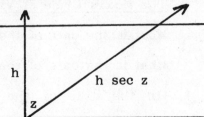

1. When starlight passes through the Earth's
    atmosphere some of the light gets absorbed
    and scattered, so that we observe the star
    to be fainter than it would be if we
    observed it from above the atmosphere.
    This phenomenon is called <u>atmospheric</u>
    <u>extinction</u>. Even if we see a star
    overhead, its light has undergone some
    weakening.   We can treat the problem in a simple way if we imagine
    the atmosphere to consist of parallel layers of decreasing density
    as we go upwards from the Earth's surface. If we have a layer of
    thickness h, and we observe a star whose zenith distance is z,
    the starlight has passed through a length h sec z of the
    atmosphere. (The trigonometrical function sec z is just 1/cos z.)
    If Δm denotes the increase in the magnitude of a star due to
    atmospheric extinction when it is seen overhead (z = 0°), the
    magnitude increases at an angle z is Δm sec z. This result
    holds good if we have a series of layers of various thicknesses,
    and more generally, for a continuously variable density atmos-
    phere.   Modern instruments can make very accurate measurements;
    for naked-eye observing the extinction does not change by more
    than 0.1 magnitude for zenith distances less than about 45°.
    Thus, to see the effect with ease one must use stars which are
    rather close to the horizon.

2. (a) Observe a southern star, as far south as possible.

    (b) Compare your star with some of the fainter stars which you can
    see nearly overhead (within about 20° of your zenith if possible).
    Select one faint star whose brightness appears to be about the
    same as the brightness of the southern star.

    (c) Record to an accuracy of about 5 minutes the time of your
    observation.   (If a sidereal clock is available, record the
    sidereal time rather than the civil time. If not, you must
    calculate the sidereal time from the civil time.)

    (d) Use <u>Norton's Star Atlas</u> to identify the two stars. (If you
    have difficulty with this, ask your instructor for help.) Record
    the star names in your notebook.

(e) Find your stars in the Bright Star Catalog and record their visual (i.e. V magnitudes) in your notebook.

3. The hour angle H of a star is the time since the star crossed the meridian. It is given by

    H = Sidereal Time - Right Ascension

    Calculate H (in hours and minutes) for your southern star. (If the star has not yet crossed the meridian, you will get a value of H which is between 12 hours and 24 hours. If necessary, add 24 hours to the sidereal time before doing the calculation for H.)

4. (a) If H is between about $23^h 20^m$ and $0^h 40^m$ then you may assume that the star is on the meridian, to sufficient accuracy. The zenith distance of the star is

    $$z = \phi - \delta$$

    where ϕ is the latitude of the observer (40° for Boulder) and δ is the declination of the star (δ is negative for southern stars).

    (b) If H is outside the above limits you must calculate z. The formula is

    $$\cos z = \sin \phi \sin \delta + \cos \phi \cos \delta \cos H$$

    where H must be expressed in degrees (1 hour = 15°).

5. If the two stars appear to be equally bright, then the difference of their true visual magnitudes outside the atmosphere (as given in the Bright Star Catalog) is equal to the difference in the extinction at the two zenith distances. Let

    V = true brightness of southern star at zenith distance z

    V' = true brightness of comparison star at zenith distance z'

Then

    $$V' - V = \Delta m [\sec z - 0.01 (\sec z - 1)^2 - \sec z']$$

$\Delta m \sec z$ is the extinction of the southern star and $\Delta m \sec z'$ the extinction of the star near the zenith, and we have added a small correction term inside the bracket to make allowance for curvature of the Earth's atmospheric layers. This correction is recommended for zenith distances greater than about 70°. If the comparison

LABORATORY EXERCISE XI

star is within about 18° of the zenith you can assume that sec z' = 1: otherwise it should be calculated from the sidereal time and the coordinates of the star.

Calculate Δm from your data. The result is the extinction expressed in magnitudes per air mass. (1 air mass is the amount of atmosphere vertically over the observer. Neglect the fact that Boulder is 1650 m above sea level.)

6. Suggested stars:

Early Fall: γ Gruis
Late Fall: one of the stars of Orion rising
                  β Eri
Early Spring: τ Pup, λ Vel, α Col, β Col.

1. Simplifying a suggestion of C.M. Snell and A.M. Heiser, Pub. Astron. Soc. Pacific 80, 336-338, 1968, we recommend using the formula

$$\text{Air mass} = \sec z - 0.010 (\sec z - 1)^2$$

for the air mass at zenith distance z. Snell and Heiser also included $(\sec z - 1)^4$ and $(\sec z - 1)^6$ terms. These can be omitted when working at the low accuracy used in our project. R. H. Garstang chose the coefficient 0.010 to give approximate agreement with the air masses given by Snell and Heiser (their Table I, column 3) and by C.W. Allen, Astrophysical Quantities, 3rd Edition, 1973, p. 125. The above formula gives an accuracy of 0.1 air mass for angles up to z = 86°, and of course it is much more accurate for smaller angles.

2. R.H. Garstang made the following comparisons on various nights:

Star	Zenith Distance	Comparison star	Estimated mag of star	True mag	Extinction mag/air mass
γ Gru	77.5	ϵ Lyr	4.0	3.0	0.29
β Ori	84.9	β Peg	2.4	0.1	0.25
β Eri	81.5	λ Peg	3.9	2.8	0.20
τ Pup	80.0	(μ Gem 2.9) (ϵ Gem 3.3)	3.1	2.3	0.18
λ Vel	83.5	(μ Gem 2.9) (ϵ Gem 3.3)	3.1	2.2	0.12
α Col	76.4	δ Gem	3.5	2.6	0.20
β Col	77.4	γ Tau	3.7	3.1	0.24
α Vir	82.0	ϵ Leo	3.0	1.0	0.34 (hazy)

We used an air mass of 1.1 for the comparison stars if they were far enough from the zenith to require it.

3. In early Fall there are very few stars suitable for this project. γ Gruis is best, apparently, and there are some in Sagittarius which could be used in the early evening in September. Fomalhaut is <u>not</u> recommended; there are no suitable comparison stars, and the zenith distance when it is near the meridian is not large enough.

4. In the second half of the Fall semester we recommend using a star of Orion as Orion rises in the east. In this case one has to calculate with some care the zenith distance of the chosen star from the time of observation. We suggest using stars in Pegasus for comparison.

5. In early Spring semester we suggest τ Puppis, α Columbae, β Columbae and λ Velorum. (It is hard to believe the extinction 0.12 obtained above for λ Velorum.) Later, it is possible to use α Vir.

ANSWERS TO QUESTIONS IN TEXT

Chapter 1: The Universe: An Overview

1. Because new instruments and the ability to get above the earth's atmosphere have enabled us to observe in parts of the spectrum other than the visible.

2. 3×10^8 m/s; 3×10^{10} cm/s.

3. 2.5 seconds. There is also a slight delay of three-tenths of a second in the land lines on earth.

4. 2h 45 min = 165 min × 60 sec/min = 9900 sec
   9.9×10^3 sec × 3×10^5 km/sec = 2.97×10^9 km
   2.97 km × 10^9 km × (1AU)/(1.50×10^{11} m/AU) × 10^{-3} km/m = 20 AU.

5. 4×10^7 years.

6. 2×10^{12} years.

7. d, b, a, e, c, f.

8. 1 mm to 1 km. 1 part in 10^{18} of total range.

9. Perhaps the earth to the moon = 4×10^5 km; earth to Jupiter = almost 10^9 km.

10. (a) 4.642×10^3; 7×10^4; 3.47×10^1
    (b) 2.54×10^{-1}; 4.6×10^{-3}; 1.0243×10^{-1}
    (c) 2,540,000; 200.4.

11. (a) 6.5×10^5
    (b) 6.5×10^8
    (c) 1.25×10^{11}
    (d) 5.6×10^{11}.

12. 3×10^5 years/1.5×10^{10} years = 2×10^{-5} = 0.002%.

Chapter 2: The Development of Astronomy

1. The planet must be going faster on its epicycle than the epicycle is going on its deferent in order for the net velocity to be backwards.

2. (a) Ptolemaic: The sun would be orbiting earth inside our orbit, and we would see the earth go through all phases, including full.
   (b) Copernican: The earth would be an inner planet, and would go through only crescent and gibbous phases, never appearing full.

3. A Saturnian would see Pluto, an outer planet much farther from the sun than it is, go through only gibbous and full phases in both Ptolemaic and Copernican systems.

4. At the time of Copernicus, there was no observational evidence

that the heliocentric system was better than a suitably adjusted earth-centered system. Copernicus did not obtain new observations; the notion that observations were as important as we now think they are was not held. It was really the discovery by Kepler that elliptical orbits gave more accurate fits to the data than circular orbits that gave the edge to the Copernican theory. The discoveries of the phases of the planets and the moons of Jupiter by Galileo were also strong points in favor of the heliocentric theory. The later proof by Foucault, with his pendulum, that the earth was indeed moving in space, and the discovery by Bessel of stellar parallaxes, provided final proof that Prolemy was wrong.

5. It is possible that some astronomical alignments could have resulted from chance, so first one has to show that there are too many astronomical alignments for them to be accidental. If this can be established, then it is clear that the people who constructed Stonehenge knew about the moving positions of things in the sky. They were at least sophisticated enough to know about phenomena that can be predicted with Stonehenge. If they indeed could predict eclipses, that means that they were considerably more sophisticated than they had to be to predict merely the solstices or even heliacal risings.

6. Mayan alignments at Chichen Itzá and Tikal, and astronomical alignments at the Bighorn medicine wheel.

Chapter 3: The Origin of Modern Astronomy

1. Phases of Venus: agreed with Copernican theory, since in the Ptolemaic theory Venus would appear only as a crescent;
Size of Venus while it goes through phases: agreed with Copernican theory, since in Ptolemaic theory Venus would always be the same distance from the earth and would thus always be the same size;
Moons of Jupiter: finding bodies that orbited a body other than the earth supported the idea that planets could also orbit a body other than the earth;
Sunspots: showed that the sun was not "perfect," and that Aristotle and Ptolemy hadn't known everything.

2. The variation in size of Venus was not accounted for by the Ptolemaic system, whereas it followed naturally in the Copernican system from the fact that both earth and Venus were orbiting the sun, and thus that the distance of Venus from the earth could vary greatly.

3. Yes, because a circle is a special case of an ellipse. The orbits of most of the planets are very nearly circular.

4. The earth is moving fastest when it is closest to the sun, following Kepler's second law. (This actually occurs during northern hemisphere winter.)

5. $P^2 = a^3 = (0.4)^3 = 0.16 \times 0.4 = 0.064$, therefore $P = \sqrt{0.064} = 0.25$ year.

6. Using Newton's form of Kepler's third law, $p^2 = a^3/9 = 1/9$, therefore $P = 1/3$ year.

7. In greels and fleels, $p^2 = a^3$, therefore $a^3 = 8^2 = 64$ and $a = 4$ greels.

8. By Kepler's third law, $p^2 = a^3$, so the period will increase by the square root of 2 cubed $= \sqrt{8} = 2\sqrt{2} = 2.8$ times longer. Since the period of a synchronous orbit is 24 hours, the new period would be about 70 hours.

9. (a) From Appendix 3:

	$a^3 \quad / \quad p^2$
Mercury	$0.3871^3/.24084^2 = 1.00002$
Venus	$0.7233^3/.61515^2 = 0.99998$
Mars	$1.5237^3/1.8808^2 = 1.00003$
Jupiter	$5.2028^3/11.862^2 = 1.0009$
Saturn	$9.5388^3/29.456^2 = 1.0003$
Uranus	$19.1914^3/84.07^2 = 1.00008$
Neptune	$30.0611^3/164.81^2 = 1.0001$

(b) Pluto's constants are poorly known. Only Jupiter's and Saturn's values show a much greater disagreement with 1 for the value of a^3/P^2 than those of the other planets.
(b) $p^2 = 4\pi^2 a^3 \div G(m_{sun} + m_{Jupiter})$. From Appendix 2, the mass of the sun $= 1.9891 \times 10^{33}$ g. From Appendices 3a and 2, the mass of Jupiter $= 317.89 \times mass_{earth} = 317.89 \times 5.9742 \times 10^{27}$ g $= 1.9 \times 10^{30}$ g. So the mass of Jupiter is 0.001 the mass of the sun, which is just enough to compensate for the deviation calculated in part (a) compared with ignoring the mass of Jupiter. Saturn's mass is about 1/3 that of Jupiter, so Saturn verifies the formula, too.

10. Kepler was working empirically, with numbers only, while Newton proved the laws from general principles.

Chapter 4: Light and Telescopes

1. No chromatic aberration; can make a larger mirror since the glass doesn't have to be perfect all the way through, and the mirror can be supported all the way across the back.

2. Atmospheric seeing.

3. Many stars, galaxies, and nebulae are visible from only one hemisphere. Many of the most interesting sources, including the center of our galaxy, rise higher in the sky and are up longer in the southern hemisphere. The Magellanic Clouds, the nearest galaxies to us, are visible only from the southern hemisphere.

4. Many clear nights, good seeing, high altitude, dry climate.

5. 1.5 magnitudes.

6. $(d_1/d_2)^2 = (1/4)^2 = .0625 = 6\%$

7. The wavelength of blue light is shorter, so smaller angles can be resolved.

8. For light of 5000 A, $(2 \times 10^{-3}) \times (5 \times 10^3 \text{ A})/(20 \text{ cm}) = 1/2$ arc sec for green light.

9. (a) reflecting; (b) both, though the situation arises most often for a spherical reflecting mirror; (c) reflecting; (d) both; (e) refracting; (f) reflecting; (g) both, for a large enough reflector.

10. Segmented mirror or multiple mirror.

11. The energy in an infrared photon is too small to expose a normal photographic emulsion.

12. The 30 arc min field of the 4-m telescope is 1/14th the diameter of the 7° field of the Schmidt and thus 1/300th the area; 900 × 300 = 270,000 pairs of plates required.

13. Ultraviolet radiation is absorbed at altitudes higher than those that balloons reach.

14. The atmosphere scatters optical but not radio radiation from the sun.

15. Large collecting area for less cost than single mirror of that size. It has 80% of the collecting area of the 5-meter telescope, so it wouldn't be quite as sensitive.

16. The radiation path is similar. For optical photographs you can use film to make an image; radio images must be built up point by point unless an interferometer is used. In the optical case, seeing limits resolution; for radio telescopes, the size of the telescope (by phenomenon of diffraction) is a limiting factor.

17. For a telescope in orbit there are no atmospheric effects; that is, all wavelengths can be observed and there are no seeing effects.

18. It is much less expensive to use a telescope on earth than in space; very large and state-of-the-art equipment can usually only be used on earth.

19. IUE is a 45-cm telescope currently in orbit, and HST is a 2.4-m telescope scheduled to be launched in 1988. IUE is a spectrographic spacecraft, with an image available only for object acquisition. HST will have a variety of imaging and spectrographic devices, and will have much higher resolution than can be attained from the ground. IUE is in synchronous orbit; ST, to be

launched by a space shuttle, will orbit the earth in about 90 minutes. IUE uses vidicons; HST will have CCD's on board.

20. The sky background is darker from space, and ST will be able to concentrate the light from distant objects in a smaller region of the image plane and thus make the image of a faint star stand out.

21. 800 km/hr × 1 hr/60 min × 1 min/60 sec = .222 km/sec
$\Delta\lambda/6600A = (2 \times 10^{-1} km/s)/(3 \times 10^5 km/s)$
$\Delta\lambda = 6600A \times (7.4 \times 10^{-7}) = .005A$
$\lambda = 6600.005A$.

22. $\Delta\lambda/6563A = (1000 km/s)/(3 \times 10^5 km/s)$
$\Delta\lambda = 6563A \times .0033 = 21.9A$
$\lambda = 6585A$.

23. $\Delta\lambda = 2A$
$2A/5250A = \underline{v}/(3 \times 10^5 km/s)$ $\underline{v} = 114 km/s$.

Chapter 5: The Sky and the Calendar

1. In a room, the separation of our eyes is large enough with respect to the distance to an object that we can see a parallax and translate it to a third dimension.

2. The Big Dipper is an asterism because it is only part of the larger constellation Ursa Major (the Big Bear).

3. No: the IAU divided the whole sky into 88 constellations.

4. No, they are sufficiently far apart, and no connections can be seen between them. Only a real clustering leads to the conclusion of this question, and even then, individual objects may be at different distances.

5. "Seeing" is image motion and "twinkling" is a variation in brightness.

6. The stars are likely to twinkle more near the horizon because we are looking obliquely though the atmosphere and are seeing through more air. Also, the air is closer to the ground and so is more subject to turbulence caused by heating from the ground.

7. Uranus would twinkle more than Venus because its disk appears smaller.

8. The north celestial pole is fixed in our sky, and stars near enough to it move in circles that are always above the horizon. Stars near enough to the south celestial pole move in circles that never reach up to our horizon.

9. The question provides practice in using the appendices, and calls attention to them Sirius has r.a. = 6^h, dec. $-16°$, and Canopus

has r.a. 6^h, dec. $-52°$; the declinations are sufficiently different that the stars are far apart.

10. $14^h14.8^m$.

11. (a) 12^h; (b) 0; (c) 0 (neglecting precession); (d) 0 (neglecting precession).

12. (a) The declination of the sun varies from 0° to +23.5° to 0° to −23.5° to 0° (starting on the first day of spring).
(b) The right ascension of the sun increases from day to day becasue the earth must rotate a little bit extra past a sidereal day for the sun to reach the meridian.

13. Every 15° corresponds to 1^h, 1/24 of the earth's rotation period.

14. 6 hours.

15. 60 min/hr × 24 hr/d ° 365 = 4 min/d.

16. One.

17. At 6 p.m. New York time on October 1st, we can look at the time zones in Figure 5-21 and see that it gets earlier hour by hour as we go west across the United States and the Pacific. As we cross the International Date Line, we add one day. Thus it is 10 hours earlier than New York counting time zones, making it 8 a.m., but 1 day later, making it October 2nd. We can see the same thing by counting time zones eastward, in which case we do not cross the International Date Line but we do obviously pass midnight.

18. LA is time zone +8 and China is time zone −8, so it is 16 hours later in China, which is equivalent to 8 hours earlier on the next day. At noon in LA, it is thus 4 a.m. the next day in China.

19. With an equatorial mount, a telescope can track the stars with a motion around only one axis.

20. With the availability of inexpensive computers to calculate and control the motions around two axes, altazimuth mounts have become practical and less expensive.

Chapter 6: The Structure and Origin of Our Solar System

1. Full moon from earth = new earth from moon; half moon from earth = half earth from moon.

2. Not significantly different with respect to the stars.

3. It would always stay in the same place. The stars and sun would move around the sky once per earth month (or lunar day).

ANSWERS TO QUESTIONS IN TEXT

4. (a) The moon's orbit is not in the same plane as the earth's.
    (b) There is an eclipse when the moon passes through the plane of the earth's orbit at new or full moon. If this happens, two weeks later it is still close enough to the plane to be partially eclipsed.

5. The earth would eclipse the sun. The earth would appear so large that not only the solar photosphere but also the corona would be covered. (At a solar eclipse observed from earth, the moon covers the solar photosphere precisely, leaving the corona in view.) The earth's atmosphere around the terrestrial limb would appear reddish, whereas the moon has no atmosphere. If there is a total eclipse as seen from earth, then the solar eclipse as seen from the moon would be seen over the entire half of the moon facing the earth. In contrast, we see a solar eclipse only from a narrow band on the earth's surface.

6. We would see the earth as a black dot in transit between us and the sun. We would see a transit rather than an eclipse because the earth's angular size from Mars is too small to cover the sun.

7. The earth's atmosphere scatters light from the photosphere. This scattered light hides the corona.

8. The photosphere is too bright. The whole photosphere is 1 million times brighter than the corona so even only 1% of the photosphere is 10,000 times too bright.

9. The chromosphere is pinkish because of hydrogen, which glows red from its H-alpha line.

10. From the ground, we can see the inner and middle corona better than from space, and do it at much lower cost. From space, we can see the outer corona, map it hourly for as long as a satellite is aloft, and study coronal x-ray radiation.

11. What counts is the semimajor axis rather than an instantaneous position of an orbiting body. Thus the fact that Pluto is in such an eccentric orbit that it is much closer to the sun than its mean distance or even closer than Neptune doesn't affect the average period.

12. Once divers start to rotate they can adjust their rate of rotation by changing their distribution of mass--the moment of inertia--through motion of the arms and legs. When divers want to slow their rotation to enter the water, they extend their bodies straight. They do this at an appropriate time so that they hit the water vertically.

13. The farther one, because angular momentum = distance $\times$ velocity $\times$ mass. Velocity = circumference/period but $P^2 = d^3$ by Kepler's third law so $P = d^{3/2}$. Thus $V = 2\pi d^{-1}P \times d \times d^{-3/2} = d^{-1/2}$. Thus angular momentum $dd^{-1/2} = d^{1/2}$, so increases with distance.

14. They didn't contain enough mass to generate enough heat from

gravitational contraction to make fusion begin.

15. The ones with the strongest gravity, namely, Jupiter, Saturn, Uranus, and Neptune.

16. At the temperature of the ring, the emission of the material is strongest in the infrared (as expressed by Planck's law and Wien's law).

ANSWERS TO QUESTIONS IN TEXT

Chapter 7: Our Earth and Moon

1. The core is densest, followed by, in decreasing order, the mantle and the crust. The crust, indeed, floats on the mantle. The upper mantle and crust together form the lithosphere.

2. (a) Tides originate because of the differential pull of the moon on the earth. (The differential pull of the sun is a secondary cause.) The water is pulled away from the earth on the side nearest the moon; on the far side, the earth is pulled away from the water. The details of local tides depend on the shape of the coast, depth of channels, etc. (b) The tides would be lower if the moon were twice as far away. Tidal forces diminish with the fourth power of distance, so then solar tides would be stronger than lunar tides.

3. If r is the earth's radius and d is the earth-moon distance, then the differential force is $GMm[1/(d-r)^2 - 1/(d+r)^2]$. For d = 384,500 km and r = 6,378 km, the differential force is -4.5×10^{-13}, which is 0.07 times (7% of) the force at distance d.

4. The diagram should show that the sun, earth, and moon make a right angle. Then the solar high tide would correspond to the lunar low tide.

5. The sun.

6. See Figure 7-3. The east coast of South America can fit into the west coast of Africa; Africa can fit into Europe; India can fit into the southwest coast of Africa; the east coast of Africa can fit onto northwest Africa.

7. 37 km. (The graph only shows that it is above 30 km.)

8. In the troposphere, the temperature decreases with altitude. In the stratosphere, it increases with altitude but at a slower rate.

9. The astronauts are below the inner Van Allen belt and the communications satellites are above the outer one.

10. Volume $\propto r^3 = (6,378/1,738)^3 = 50$ times larger.

11. The moon's radius is smaller than that of earth.

12. The line of sunrise or sunset.

13. Energy is given off as radioactive material decays. This energy accumulates in the interior and then flows out gradually. Any energy from radioactive decays near the surface would flow out more readily, making the rate higher than otherwise.

14. That it is much older than the surface of the earth, since too much cratering is present to have only taken place recently.

15. The maria were covered with lava after the rest of the surface had formed.

16. The initial surfaces on the moon are better preserved than on earth as a result of erosion on earth.

17. The flows in the Sea of Fertility were more recent than those in the Sea of Serenity. Apollos 15 and 17 went near the Sea of Serenity; no Apollo mission went near the Sea of Fertility (Foecunditatis) but 2 Soviet Luna missions went there. See Figure 7-18.

18. The whole lunar interior is not molten.

19. See Section 7.2g.

20. (a) See Section 7.2b. American: Apollo had 6 moon landings. Soviets: unmanned craft including Lunokhods (rovers) and Lunas, three of which returned soil to earth.
(b) See Section 7.3. No definite American plans. Soviet plans are for an unmanned mapper in polar orbit.

Chapter 8: Mercury

1. From the equator, Mercury would approach the horizon perpendicularly. At latitudes north or south, it would approach at an inclination. The longitude of the observer affects only the time of sunset.

2. No. The back side would face all parts of the heavens during a Mercurian year.

3. It takes 2 full orbits of Mercury for the sun to return to the same point in the sky over Mercury.

4. Radar told us the speed of Mercury's rotation from the Doppler shift of the returned signal.

5. Mercury's albedo is 6%, so 0.06 erg is reflected each second from 1 erg incident.

6. Mercury is more likely covered with basalt than ice.

7. Decrease. Less radiation would be absorbed.

8. Surface structure; magnetic field; presence or absence of atmosphere.

9. Measure intensities point by point and radio back individual intensity values. Make measurements in a raster pattern (that is, back and forth, eash line being displaced, as on a tv).

10. An old crater would have more small craters superimposed.

11. Visual appearance. Erosion could have been from impacts of

meteorites or micrometeorites, solar wind, or lava.

12. Photographs of the surface (presence of craters, scarps); magnetic field; presence of and composition of atmosphere (presence of helium); surface covered with dust.

Chapter 9: Venus

1. Similarities include size, density, and presence of atmosphere. Differences include the greenhouse effect on Venus, lack of water on Venus, and amount of CO_2 in the atmosphere of Venus.

2. The CO_2 on earth is locked in the rocks. The difference may result from Venus's original (pre-greenhouse), slightly higher temperature because it is closer to the sun.

3. Its rotation is retrograde.

4. Depending on what region you watched, you would get a different rotation period if you watched for only a short time; the cloud rotation is not steady. But the average cloud rotation rate is about the same as the surface rotation rate.

5. The planet would be very cold. No optical radiation would get in, and any radiated heat would escape in the infrared.

6. The clouds don't interfere with the radio waves that we transmit to make the radar maps.

7. If the pressure on Venus is over 90 times earth's surface pressure, and we remove 96% of the atmosphere (the carbon dioxide makes up 96%), there would still be enough atmosphere left to rival the earth's in pressure.

8. The particles must be charged.

9. If the earth had been hotter, the CO_2 would not be tied up in the rocks and would be in the atmosphere, starting a large greenhouse effect.

10. The planet would cool, at least temporarily.

11. It is about the same size, mass, and density as the earth.

12. The Venera 9 showed sharp-edged, angular rocks. Venera 10 showed smoother rocks with lava or debris between them.

13. The ground-based radar observations have resolution of typically 10 or 20 km, better than satellite resolution but only over a much more limited section of Venus than the Orbiter radar observations. The Orbiter was able to map almost the entire planet (93%) with a resolution "footprint" of 100 km and get a global view, as shown in Color Plate 17. It studied the distribution of highlands and lowlands, and thus concluded that Venus

did not undergo tectonic motion and appears to be of one tectonic plate. The ground-based radar observations did map several volcanoes, the continent Terra Ishtar, and the giant valley. Ground-based resolution of 20 km on Venus has been obtained. The radar aboard Magellan, now postponed, would map nearly the whole surface of Venus with a resolution of 1 km, plus a few selected regions with a resolution of 0.1 km.

14. Radio waves penetrate clouds (which is why we can receive radio and tv signals on earth even through clouds), so give us views of Venus's surface.

15. General rolling plains cover 60% of the surface, lowlands only 16%, two continents (Ishtar, Aphrodite), mountain Maxwell on Ishtar, Beta Regio (volcanic region), and giant canyon.

16. It has the effect of a larger aperture and hence higher resolution.

17. Varying abundance of sulfur dioxide and lightning near Beta Regio and Aphrodite Terra.

18. Venus apparently has 1 equator-to-pole circulation, while earth has 3 patterns in latitude bands in each hemisphere.

Chapter 10: Mars

1. Reasonable temperatures, presence of atmosphere and solid surface, evidence for water in the polar caps and in the past on the surface.

2. Less dense if the only effect of closeness is to heat the atmosphere and allow more molecules to escape, but the total result may be very complicated.

3. $(M_{Mars}/M_{earth})/(R_{Mars}/R_{earth})^3 = 0.1/(.53)^3 = 0.1/.15 = .7$.

4. $1 \div (1/1.5)^2 = 1 \div (2/3)^2 = 9/4 = 2.25$ times more.

5. Relative to planetary radius, Olympus Mons is 5 times the height of Mauna Kea.

6. Channels that appear to have been carved by water and signs that something flowed around obstacles. Evidence for water in ice caps.

7. From the pinkish dust suspended in it.

8. The north polar cap has a residual part of water ice, under a changeable part of dry ice (carbon-dioxide ice). The south polar cap's top part is also dry ice, and there is no conclusive evidence whether or not there is a water-ice residual cap too. We know that the residual north polar cap is water ice because its temperature, measured from the Viking Orbiter, is too warm

for it to be frozen carbon dioxide.

9. Spectroscopy, occultation of radio signals from space probes, and direct sampling.

10. Mercury has the largest range, Venus is the hottest but is very steady, earth and Mars have comparable ranges but Mars is colder.

11. As described in the text, the three biological experiments gave evidence that at first could be interpreted in terms of life but eventually gave results that matched strange soil chemistry. See Box 10.1 for details. The chemical experiment, the mass spectrometer, found no sign of organic material, a strong point against the existence of life on Mars.

12. Bacteria would do well; the equipment was not designed for large life forms like humans, though we certainly could incorporate the nutrients, give off carbon dioxide, and show signs of organic molecules in a mass spectrometer. (I wouldn't want to be pyrolized, though.)

13. (a) Seismographic experiments, photography of surface, and study of Martian "weather." (b) Spectroscopy, mapping of the surface, and temperature measurements.

14. They do not have enough mass to have had enough gravity to make them round. They may be pieces of rock left from meteoroid impacts.

Chapter 11:   Jupiter

1. Jupiter is larger and has a higher albedo.

2. Differential rotation will cause the line to bow.

3. Probably the Great Red Spot has a source of heat below, deep in the atmosphere. Storms on earth do not have such sources of heat. Also, Jupiter has no continents or other structure to break up the storm. The fact that Jupiter's clouds radiate less efficiently than the clouds of earth may also be a factor.

4. (a) First knew from radio bursts detected on earth.
(b) Magnetic field is more extensive than we had thought. Also, we learned about its shape and how its shape changes with time.

5. Jupiter can be as much as 6 A.U. from the earth when it is on the other side of the sun from the earth (5 A.U. from the sun for Jupiter plus 1 A.U. from the sun for the earth), and can be as near as 4 A.U. (5 A.U. - 1 A.U.), rounding the radius of its orbit from 5.2 to 5 A.U. for convenience. Thus when Jupiter is relatively far, the light from eclipses could take an extra time to travel equal to $(2 \text{ A.U.})/c = (3 \times 10^8 \text{ km})/(3 \times 10^5 \text{ km/s}) = 10^3 \text{ s} = 15 \text{ min}$, compared with the schedule of the light's arrival from eclipses set up when Jupiter is relatively near.

Roemer actually made his deduction of the speed of light from the fact that the eclipses were later when Jupiter was moving away than they were when Jupiter was approaching the earth, rather than from absolute timing.

6. Jupiter has an internal heat source and emits more energy than it gets from the sun, as verified by infrared observations first from the ground and now from spacecraft including the Voyagers. Also, Jupiter's chemical composition is 90% hydrogen and 9% helium, similar to the chemical compositions of the stars but very different from the compositions of the terrestrial planets. Nevertheless, Jupiter has too little mass by a considerable factor to be a star.

7. No atmospheric seeing problems; closer views.

8. Magnetic field, broadcasting a radio signal for radio occultation studies, having its path tracked to measure masses for Jupiter and its moons. The text mentions that the Voyagers determined temperatures in interplanetary space near Jupiter. The Voyagers also carried many other instruments, not specifically discussed in the text, such as infrared and ultraviolet spectrometers, a cosmic-ray detector, a radio-astronomy antenna, and a charged-particle detector.

9. Interior of Jupiter may have solid core, unlike earth's molten core. Jupiter is almost entirely hydrogen and helium, unlike heavy elements in all parts of earth. Jupiter has liquid metallic hydrogen and liquid molecular hydrogen zones, unlike earth.

10. The blob goes roughly a tenth of the way around between frames, an interval of 22 hrs. The Red Spot measures 14,000 km × 30,000 km, so it is roughly $(14+30+14+30) \times 10^3$ km = 10^5 km around (if we think of the Red Spot as a rectangle and add the lengths of each of the four sides). If we approximate the circumference of the Red Spot, instead, as a circle with an intermediate radius of about 20,000 km, its circumference is then $2\pi r = 6 \times 20,000$ km = 10^5 km, the same answer. Thus the blob went about 10^4 km in 22 hours, so goes roughly 500 km/hr. If one estimates that the blob goes 1/20 of the way around between frames instead of 1/10 of the way, the value is reduced by a factor of 2. Note that neither of these numbers is the 6-day rotation period of the Great Red Spot itself.

11. Among the Galilean satellites, Europa, Ganymede, and Callisto appear to be icy because of the extreme coldness at that distance from the sun. Io is not icy because of its internal heat and volcanoes.

12. The volcanoes of Io erupt higher and for a longer duration than volcanoes on earth. The material that comes out in Io's volcanoes is sulfur and its compounds. Note that eruptions of Hawaii's Kilauea volcano, with lava being ejected, have lasted several months continuously. During a recent 5-year interval,

Kilauea's eruptions occurred much of the time. Lava erupted on one occasion in a fountain 0.5 km high. Kilauea's eruptions are more similar to Io's long-lived eruptions than the isolated eruptions of Mt. St. Helens.

13. Callisto is so covered with craters that it must be the oldest of the surfaces of the three. Io, recovered time and time again by sulfur lava flows, has the youngest surface. The moon, with many craters but with maria covered with lava, has a surface of intermediate age. The comparison of Callisto and the moon depends on the assumption of a uniform cratering rate near Jupiter and near earth.

14. The proportion of Jupiter's rings is shown on Figure 11-24, and the orbits of the moons can be added.

15. (a) The gravity-assist method sends a spacecraft near Jupiter (or another massive body) such that its orbit is changed in the gravitational interaction, both altering the direction and adding energy (and thus velocity). (b) The method has been used to send the Voyagers from Jupiter to Saturn, for example. It has been also used in Mariner 10 to Mercury and Venus, in Pioneers 10 and 11, and in other space missions.

16. It is so much larger than the Voyager telescopes that it is inherently capable of finer angular resolution before diffraction limits it; its mounting is esecially free of jitter.

17. Since π radians = 180°, 0.1 arc sec = π radians/(180x60x10) = 5 x 10^{-7} radians. The earth-Jupiter distance can be 5.2-1 = 4.2 A.U. Thus 4.2 A.U. x (1.5 x 10^8 km/A.U.) x (5 x 10^{-7}) = 315 km, roughly the distance from New York to Boston. Can also scale from the sun, since the text states that 1 arc sec on the sun = 700 km on the solar surface. Jupiter is 4 times farther away than the sun, but the resolution is 10 times better, and 400 x 4/10 = 280 km, in rough agreement.

18. Io is essentially the same distance from earth as Jupiter. Io is 3632 km in diameter, about 1/4 the earth's diameter. From Color Plate 23, you can see that the central volcano is about 1/8 Io's diameter, or 400 km, so we may barely be able to see individual volcanoes.

Chapter 12: Saturn

1. Similarities: general structure and composition, internal energy source, density, rapid rotation, magnetic field, large number of moons, and bands on atmosphere.

2. 3 arc min; about 10 times smaller than from earth.

3. Jupiter and Saturn are separated by 4 A.U. at their closest approach, about the same as earth-Jupiter distance, so size is comparable, about 40 arc sec.

ANSWERS TO QUESTIONS IN TEXT

4. The Roche limit is the distance within which tidal forces prevent a body from coalescing. Thus the material within Saturn's Roche limit formed rings instead of moons.

5. Titan's orbit has a semimajor axis of 1,221,600 km, and the B ring is roughly 100,000 km in radius. The tidal force from Saturn results from Saturn's gravity, and is concentrated at Saturn's center of mass. Titan is 5,150 km in diameter.
    (a) $1/94,850^2 - 1/105,150^2 = 2 \times 10^{-11}$.
    (b) $1/216,450^2 - 1/226,750^2 = 2 \times 10^{-12}$, 10 times smaller.

6. The orbits are all outside the rings; the scale of the rings can be measured on the photos and compared with Saturn's diameter.

7. The chapter discussed first the closeup and odd-angle views from Pioneer 11, and then the revolution in our understanding of Saturn's rings, moons, and atmosphere from Voyager observations.

8. They have solid surfaces and atmospheres.

9. Nobody is sure why there are so many rings (ringlets). Possibly disturbance from small satellites. Resonances from outer moons may also play a part, but cannot account for all the ringlets. Shepherding satellites apparently hold the material in the ringlets.

10. Small particles scatter light forward, so rings made of small particles would be expected to look bright when looking through them at the sun but dark from the back.

11. There is material in Cassini's division.

12. Spokes are radial features that form and die as the rings rotate. They may well result from electrostatic forces levitating dust above the rings.

13. Titan: completely covered with clouds, dense smog with hydrocarbons, atmosphere contains mostly nitrogen, surface pressure is 1.5 times earth's, methane cycle may lead to liquid methane on the surface.

14. Methane is near its triple point, so it can form clouds, rain, oceans, and ice; ethane may be mixed in with methane in the oceans, according to the observations.

15. Mimas: huge impact crater, surface saturated with craters, canyon may have resulted from focusing of impact forces from formation of the giant crater;
    Enceladus: smooth regions and cratered regions, long linear grooves;
    Hyperion: irregular shape, many craters;
    other examples in text.

ANSWERS TO QUESTIONS IN TEXT

Chapter 13: Uranus

1. Axis lies close to the plane of the solar system. No seasons or variable heating at different latitudes that would lead to weather effects.

2. Uranus's moons are comparable to Saturn's Tethys and Rhea and our moon; Neptune's Triton is comparable to Io and Europa, and only somewhat smaller than Mercury.

3. If a star is occulted in an off-center passage, we know that the planet is at least as large as the angular distance between the center of the planet and the position of the star. Also, by timing the occultation we get a lower limit to the planet's size.

4. Occultation of a star.

5. Albedo, variation in width and albedo with position, one new narrow ring, a new broad band.

6. The clouds on Uranus are drawn out in latitude, so rotation around Uranus's axis seems more important than the orientation of its axis with respect to the sun.

7. Methane in Uranus's atmosphere absorbs the orange and red, leaving mostly blue-green.

8. Sunlight splits hydrogen molecules into protons and electrons. The electrons somehow acquire extra energy and collide with remaining hydrogen molecules to make the electroglow.

9. Frontlighted: narrow, dark rings. Backlighted: brighter bands of material. The forward scattering causes the backlighted bands to show, so there is dust but less dust than on Jupiter.

10. Uranus's magnetosphere is about 50 times stronger than earth's. It is tipped 60° and is off-center. The earth's magnetosphere is much less tipped and is centered on the earth.

11. Voyager flew by so quickly that the moons did not have time to rotate and Voyager's path did not bend enough to give other views.

12. See section 13.4d.

13. Ariel has the youngest surface because it shows so much geologic activity.

14. Umbriel and Oberon apparently have the oldest surface because of all the craters visible.

15. The minor satellites have lower albedos than the major satellites.

ANSWERS TO QUESTIONS IN TEXT

Chapter 14: Neptune and Pluto

1. Among the giant planets, Jupiter, Saturn, and Neptune.

2. First seen by Galileo in 1613, about 375 years ago, and discovered in 1846, about 140 years ago; 165-year period. About 85% of its orbit since it was discovered and 226% since it was first seen.

3. Because Voyager 2 will fly so close to Triton in 1989.

4. About 20%.

5. The eccentricity and inclination of the orbit. Its small size, coupled with its location beyond the giants. Its extremely low density.

6. Uranus is 58 K, so if a 10th planet is beyond Pluto, 30 K is a reasonable estimate. At such a low temperature, a body would give off most radiation in the short-wavelength end of the radio spectrum, though it would still be reasonably bright in the infrared; best to look at IRAS or other infrared data, since it is too difficult to map the whole sky at 1 mm.

7. It is much smaller (occultation, speckle interferometry), and of much lower mass (derived from Charon).

8. From the equation given in Box 3.3, we can substitute the measured values for the period P and the semimajor axis a and derive the sum of the masses, given the value of G in Appendix 2. $(m_1+m_2) = 4\pi^2 a^3/GP^2 = (4)(10)(17,000 \text{ km} \times 10^3 \text{ m/km})^3 \div (6.672 \times 10^{-11} \text{ m}^3/\text{kg}\cdot \text{s}^2)(6.4 \text{ days} \times 24 \text{ hr/day} \times 3600 \text{ s/hr})^2 = 9.5 \times 10^{21} \text{ kg}$. The earth's mass, from Appendix 2, is $6 \times 10^{27} \text{ gm} = 6 \times 10^{24} \text{ kg}$. This makes Charon's mass 1/600 or so the mass of the earth. The calculation could also be done in terms of earth values.

ANSWERS TO QUESTIONS IN TEXT

Chapter 15: Halley's and Other Comets

1. Heading towards the sun.

2. Because it lists all the fuzzy or unusual objects in the sky that might at first glance be confused with a comet.

3. Not necessarily. Comets can come from any direction in the comet cloud.

4. About 1,000 times as distant.

5. Indirectly from the radiation of the sun.

6. When they pass close to the sun, more material is evaporated away from the nucleus.

7. Nucleus.

8. It required ultraviolet observations, which can only be made from above the earth's atmosphere.

9. Structure given by variations in the solar wind; ion tail.

10. $a^3 = p^2 = 76^2 = 5776$, therefore a = 18 A.U. Major axis = 2a = 36 A.U. The orbit is very elliptical (ellipticity + 0.967), and the sun (out of focus) is near an end of the orbit. Draw the ellipse with one end about half-way between the earth's orbit and the sun (0.6 A.U.), and the other end about 36 A.U. out, between the orbits of Neptune and Pluto. (A common student error is to center the orbit on the sun, instead of putting the sun at one focus of the ellipse).

11. $a^3 = p^2 = 80,000^2 = 6.4 \times 10^9$, therefore a = 1,850 A.U., about 50 times farther out than Pluto's orbit.

12. Magnetic field measurements, found magnetic field draping, and discovered heavy ions carried 4 million km outward by the solar wind.

13. The earth was on the opposite side of the sun from the comet when the comet passed perihelion.

14. The Vegas and Giotto imaged Halley's jets.

15. Halley's Comet is potato-shaped, about 15 km × 10 km × 10 km, and covered with a dark crust.

16. The dust particles are rich in H, C, N and O. Many are small, and there were more of relatively low mass than expected.

17. The maximum magnetic field was 65 nanoteslas. A region of no magnetic field was found, as was a bow shock.

18. The discoveries endorsed Whipple's model, though no one realized

that the activity would be so concentrated in jets.

19. IUE found variations in gas output with a time scale of hours.

20. They are longer exposures than the 1/30 sec response time of the eye and were taken with instruments with larger apertures.

Chapter 16: Meteorites and Asteroids

1. Many meteorites may be chips off asteroids. Other meteorites, particularly those in showers, seem to be associated with defunct comets.

2. They burn up in the atmosphere.

3. It depends on the extent of the debris in space.

4. They provide information on the original composition of the solar system.

5. The largest asteroids are comparable in scale with the moons of Saturn (except for Titan); some asteroids and some moons are, similarly, differentiated.

6. If the starlight is shut off abruptly, then there is no atmosphere. (Actually, diffraction would cause the starlight to fluctuate in a known manner, so we could still tell; diffraction is not discussed in the book.) If the starlight dimmed gradually or irregularly, that would imply that an atmosphere was present.

7. Chiron may show that there is no sharp distinction between asteroids and comets. Chiron's elliptical orbit may be typical of orbits that cause asteroids to collide, leading to meteoroids (though most such collisions take place in the asteroid belt).

8. Smallest: Leda, with diameter of perhaps 10 km
   Largest: Ganymede, with diameter of 5276 km.
   Other small satellites have also been discovered, and the smallest size is a function of our abilities.

Chapter 17: Life in the Universe

1. They show that simple fundamental organic molecules can be synthesized in a relatively easy fashion.

2. Different mixtures of chemicals, reflecting current theory of the earth's primitive atmosphere, and different kinds of excitation (sparks, uv).

3. A large proper motion probably just means the star is close to us. The presence of planets would be indicated by irregularities in the proper motion.

4. Barnard's star may have a wobble in its proper motion that could be explained by one or two giant-sized planets. However, not all observers agree that the data require that the wobble is present. They feel that the wobble reported merely results from uncertainties in the observations.

5. The Hubble Space Telescope will have higher resolution by a factor of 7 than ground-based telescopes, and hence will be able to detect finer wobbles in the motions of stars than we have been able to detect up to now. Also, since its resolution is higher, it is not inconceivable that it could observe a planet directly, separating it from the glare of the associated star.

6. The radial-velocity method does not depend on the distance to a star.

7. 10^{-4}. 10^7 such stars.

8. Each person will have his or her own calculations.

9. The current value is 4.8 billion people. In reading the numbers, note that the top lines of the message give 1-10: 001, 010, 011, 100, 101, 110, 111, 1 in the left column and 000 in the right column, 1 in the left column and 001 in the right column, and 1 in the left column and 010 in the right column. Then notice that the numbers for population are:

    110110X
    111111
    111011
    110111
    111111
    11.

    where X is a spacer. See "The Search for Extraterrestrial Intelligence" by Carl Sagan and Frank Drake in the May 1975 Scientific American.

10. The telescope is 1000 feet across, and the units given are in multiples of 12.6 cm, the wavelength of the transmission. To read the binary numbers, you have to hold the page upside don, which gives

    111110X
    100101

    where X is a spacer. see "The Search for Extraterrestrial Intelligence" by Carl Sagan and Frank Drake in the May 1975 Scientific American.

11. Massive stars stay on the main sequence for relatively short times, probably too short for life to evolve.

12. Detecting radiation such as TV or radio signals from them. Neutrino messages are a second possibility, thought by most astronomers to be much less likely. Others: abundant oxygen molecules, other imbalance of molecules, finding infrared signs of a Dyson civilization.

13. Even if we could communicate with intelligent life forms around the star nearest to the sun it would take radiation carrying our messages about four years to get there.

14. Einstein's theory first explained the known discrepancy in Mercury's orbit, but later also predicted unknown effects on the positions of stars that were verified during a total eclipse. His theory proved to be more general than Newton's, and thus turned out to be the simplest explanation to explain and predict all data about gravitational phenomena. (In other words, Einstein's theory fit the requirements of Occam's razor.)

Chapter 18: Ordinary Stars

1. A star emits its own radiation, while a planet reflects radiation from its star. Stars are point sources and twinkle, while planets show disks and usually do not.

2. 2^2 = 4 times. 20^2 = 400 times.

3. shorter: 2800 A; 2^4 = 16 times the energy (assuming equal surface areas).

4. The O star's temperature is 40,000/5,800 = 7 times higher; therefore, it peaks at 1/7 the wavelength of the sun's peak = 800 angstrom.

5. The ultraviolet. No, because UV radiation does not pass our atmosphere.

6. The one that peaks at 2000 A is 5 times the temperature and gives off more energy at all wavelengths. The ratio of energy given off is 5^4 = 625 (assuming equal surface areas).

7. Each square cm radiates 2^4 = 16 times more energy. The B star's surface area is 5^2 = 25 times greater than the sun's, so the whole star radiates 16 × 25 = 400 times more.

8. Planck's law. Wien's displacement law can be derived from Planck's law.

9. $(40,000 \text{ K}/6,000 \text{ K})^5 = 6^5$ = (estimating) 40 × 40 × 6 = 1600 × 6 = 10^4 per unit surface area.

10. B, A, C.

11. (a) The Lyman series should be drawn in the ultraviolet far below the Balmer series, which is in the visible.
(b) Only recently have we been able to observe stellar spectra outside the visible, so most studies until recently have been of the Balmer.

12. Absorption lines represent diminution of continuum strength over

a narrow band of wavelength. Emission lines represent energy added to continuum (or zero) over a narrow band. See Fig. 18-6, lower right, for a continuum with an absorption line. You cannot draw absorption lines without a continuum. You can have emission lines without a continuum. Absorption lines need something to absorb into, whereas emission lines are just radiation from the gas itself.

13. (a)

```
----------5
--┬-----┬--4
  |      ↓ --3
--|--┬------2
  ↓  ----------1
```

(b) The transition from level 4 to level 2 is a vertical line drawn from level 4 to level 2. The transition from level 4 to level 3 is a vertical line drawn from level 4 to level 3.
(c) The distance from level 4 to level 2 is twice the distance from level 4 to level 3, so the ratio of the energies is 2.
(d) The energy of the transition from level 3 to level 2 is also half the energy of the transition from level 4 to level 2. The wavelength of the transition from level 3 to level 2 must thus be twice 4000 A = 8000 A, since $E = hc/\lambda$.

14. The solar surface, like that of almost all stars, shows absorption lines. A B star has stronger hydrogen lines, and the sun and other G stars have stronger H and K lines of Ca II, more iron lines, etc.

15. (a) Only O stars are hot enough to ionize helium.
(b) The solar photosphere is too cool to excite helium.

16. The Planck curves for O stars are higher at all wavelengths and peak farther to the blue than do Planck curves for G stars; the relation of Planck curves for G stars and M stars is similar to that of O and G stars. The relation of the Planck curves is most significant for a square centimeter of the surface of each type of star, but the relation of intensity is increased by the fact that O stars are bigger than G stars, which are bigger than M stars, for stars on the main sequence.

17. The solar atmosphere is sufficiently hot that molecules break apart. (Molecules do appear in the spectra of sunspots which are cooler.)

18. The results of student contests are often interesting.

19. Hα's wavelength is a constant × $(1/2^2 - 1/3^2) = 0.138888$ × constant. Hβ's wavelength is the same constant × $(1/2^2 - 1/4^2) = 0.1875$ × constant. Thus the wavelength of Hβ = $(0.1288/0.1875)$ × 6563 = 4861.

20. Surface temperature. We would determine spectral type with a spectrograph, and comparing the spectrum we measured with

standard spectra. One can also deduce spectral type by finding the color index of a star using filters and photomultipliers. (The additional dimension of luminosity classes adds a dependence on size or surface gravity.) Chemical composition is important in determining the latest spectral types (C-type and S-type).

21. (a) Hydrogen lines are strongest in A and B stars (actually between B9 and A0), and fall off in intensity in both directions. The text discusses alphabetical order by hydrogen line strength, so A star is an acceptable answer. After B stars come F, G, and M. O stars are so hot that they also have only weak hydrogen lines. The measure used for spectral types was "the ratio of Balmer lines to the intensities of a number of other lines."
(b) Ionized calcium lines peak in K stars, are strong in G stars, and fall off in both directions. Thus F stars are next, followed by A and M stars. The lines disappear, going to hotter types, by B5.
(c) Titanium oxide lines are visible only in M stars and K4–K9 stars.

22. The wavelengths of $H\varepsilon$ of hydrogen and H of Ca II are at almost the same wavelength: 3970 A; as the hydrogen lines get weaker going to cooler temperatures, the H line gets stronger. Ca K varies by itself with temperature.

23. The large stone radiates 4 times the energy. (It has 4 times the surface area.) Each square cm of the surface of each will have the same surface brightness.

24. Because only one state involved in the transition is a discrete level; the other state is "free" and can have any energy.

Chapter 19: Stellar Distances and Motions

1. 4th magnitude; 6th magnitude is the approximate limit.

2. Mars, since +0.5 is brighter than +0.9.

3. $(2.512)^3$ = 15 times.

4. 5 magnitudes = 100 times.

5. The difference is 14 − (−4) = 18 magnitudes. Fifteen magnitudes is a factor of 10^6 and the other three magnitudes make a factor of 15. Venus is thus brighter than Pluto by 1.5×10^7 times.

6. 3 magnitudes.

7. A factor of 40 is four magnitudes and a factor of 100 is 5 magnitudes. Thus a factor of 60 is about 4.5 magnitudes. 5 − 4.5 = 0.5 magnitude.

8. 6 magnitudes = 250 times.

ANSWERS TO QUESTIONS IN TEXT

9.   10,000 times = 10^4 = 10 magnitudes, so star B is 11 - 10 = 1st magnitude. Star C is 11 + 10 = 21st magnitude.

10. B is 5 magnitudes = 100 times brighter than A.   C is 7 mag = 600 times brighter than A.   C is 2 mag ($=2.5^2$) = 6 times brighter than B.

11. 3 magnitudes = 2.5^3 = 15 times brighter.

12. Parallax. In England, the driver is on the right and the passenger would see a higher speed on the dial.

13. Larger, because the same linear displacement of the earth corresponds to a larger change of angle of the star against the distant background.

14. The star with a parallax of 0.02 arc sec is 10 times farther away than a star with a parallax of 0.2 arc sec.

15. 1/0.05 arc sec = 20 pc. 20 pc × 3.26 l.y./pc = 65 l.y.

16. About 100 meters. 3 arc min.

17. 1/8 = .12 arc sec.

18. Luminosity vs. temperature.

19. For a large class of stars, temperature and luminosity have a simple relation. In particular, for stars that are fusing hydrogen to helium, stars of larger mass have surface temperatures and luminosities larger than less massive stars in a certain way.

20. A dwarf is an ordinary main-sequence star. A white dwarf, the end product of evolution of a star of about 1 solar mass, appears much fainter for its temperature than a main-sequence star would. Note: There is no simple answer in terms of colors.

21. The surfaces are the same temperature, so the one that is twice as far away has to have twice the radius.

22. The nearer star appears 100 times as bright, so the magnitude difference is 5 magnitudes.

23. It is 10 times farther from the sun than the standard 10 parsecs, so must be 100 times fainter than its absolute magnitude, which is equivalent to 5 magnitudes fainter. Since it appears at apparent magnitude 5, it must have absolute magnitude 0. From the Hertzsprung-Russell diagram, reading horizontally from absolute magnitude to the main sequence and then down, that corresponds to an A star.

24. It is 3 times farther from the sun than the standard 10 parsecs, so must be 10 times fainter than its absolute magnitude, which is equivalent to about 2.5 magnitudes fainter. It must thus be

approximately absolute magnitude 2 - 2.5 = -0.5.

25. It is 5 magnitudes or 100 times fainter than its absolute magnitude, so must be 10 times farther away than 10 pc, and is thus 100 pc away.

26. It is 4.4 magnitudes or about 60 times brighter than it would be at 10 pc, so it is about 8 times closer or 1.2 pc = 4 l.y.

27. It appears 6 magnitudes or 250 times fainter than it would be at 10 pc, so it is 16 times farther away or 160 pc = 520 l.y.

28. It is 10^5 times farther away than 10 pc, so its absolute magnitude must be 10^{10} brighter than its apparent magnitude = 25 magnitudes brighter, so it is absolute magnitude -12. It is 17 magnitudes brighter = 6 million times brighter than the sun.

29. Cooler; the spectrum is redshifted.

30. (a) That Barnard's star is relatively close to us.
     (b) Half as much.

31. Not without knowing how far it is away.

32. (a) Both stars have the same radial velocity.
     (b) The closer star, A, has the larger proper motion.

33. ST will be able to measure the positions of stars more accurately than we can now, so we can detect proper motions in shorter times.

34. Trigonometric parallax.

ANSWERS TO QUESTIONS IN TEXT

Chapter 20: Doubles, Variables, and Clusters

1. Draw a wide orbit that we view in its plane, with the two stars large enough to eclipse.

2. We find an astrometric binary from deviations of its proper motion from a straight line, and an eclipsing binary from variations in the brightness. Astrometric binaries must be relatively close for us to detect them, which is not a requirement for eclipsing binaries.

3. (a) Both dips are the same width and depth.
    (b) The dips are wider than they were in (a).

4. Using Figure 20-7, a star of 10 solar masses is 10 magnitudes or 10^4 times brighter than the sun.

5. From the H-R diagram, we can tell that the spectral type of an M2 supergiant corresponds to M=-5; on the mass-luminosity relation, this corresponds to 10 solar masses.

6. From Figure 20-7, we see that stars of 3 solar masses are about 3 magnitudes or 10 times brighter than the sun. Rough answers will do here since the horizontal scale gives equal steps of log mass.

7. (a) 10^6; (b) 10^4.

8. Most of the light comes from the brightness and most massive stars; most of the mass comes from the faintest and least massive stars.

9. Low-mass stars--there are many more of them.

10. -4 to -5, from Figure 20-13.

11. As long as they are at the same distance (as all stars in the LMC are from us), they differ by 1 magnitude = 2.5 times.

12. RR Lyrae variables are all about absolute magnitude 0.6. Its apparent magnitude is thus somewhat over 5 magnitudes less-- somewhat greater than a factor of 100 times fainter--than it would be at the standard distance of 10 pc. It is thus over 10 times farther away or somewhat more than 100 pc away.

13. 10 day period implies M=-3. Its apparent magnitude is thus 11 magnitudes fainter than its absolute magnitude, which is also true for the RR Lyrae star, making the RR Lyrae's M=11.6.

14. Draw the light curve, showing an 11-day period, with days along the horizontal axis and magnitude on the vertical axis. If it is a Cephid, measure the period, read the absolute magnitude from the period-luminosity relation, and compare the absolute magnitude with the mean apparent magnitude to find the distance.

15. 0.6, as with all other RR Lyrae stars.

16. Population I stars have similar composition to that of the sun. Galactic clusters contain Population I stars. Population II stars have abundances of heavy elements that are lower by a factor of 10 or more. Globular clusters contain Population II stars.

17. Y, since more of its stars have evolved off the main sequence.

18. <u>Main point</u>: All the stars are at the same distance. <u>Additional point</u>: By fitting the observed main sequence to a calibrated one, you can find the distance to the cluster. By observing where the stars deviate from the main sequence, you can determine the age of the cluster.

19. In the text, we described that the Hyades were about 10^9 years old, so a cluster with a turnoff farther down the main sequence, like M67, would be about 8×10^9 years old.

20. The RR Lyrae gap appears at M=15.6, which is 15 magnitudes of 10^6 times fainter than an RR Lyrae star at 10 pc; M3 is thus 10^3 times farther away, or 10^4 pc.

Chapter 21: The Sun

1. Interior: 15 million K; photosphere: 6,000 K; chromosphere: 15,000 K; corona: 2 million K; sunspots: 4,000 K; prominences: 15,000 K.

2. Most lines are iron, with lines of some other elements also present. The strongest lines are calcium, hydrogen, and sodium. Iron has lines from many of the energy levels in its complicated energy-level structure; calcium and sodium have strong lines because most of the absorbing power is concentrated in these lines, which fall in the visible. Hydrogen has strong lines beecause of its abundance.

3. We see absorption lines from the photosphere because we are observing gas against a background of a hotter continuum. We see the chromosphere at eclipse in projection against a dark sky, so it has an emission spectrum with respect to this dark background.

4. Put Hα on the right, Hβ in the center, and Hγ a shorter interval to the left. H and K are, right to the left, farther left. the sodium D lines are closer to Hα than to Hβ.

5. The chromosphere is hot enough to excite helium, and photosphere is not.

6. Start from 15 million K at the center, decrease outward through the photosphere to a temperature minimum of about 4500 K and then upward through the 15,000 K chromosphere to the 2 million K corona.

7. Filaments are prominences seen in projection against the solar disk. They appear dark against this bright background. The same objects, seen as prominences in projection against dark sky, appear bright.

8. Sunspots, flares, prominences, aurorae.

9. (a) The moon's orbit is not in the same plane as the earth's. (b) There is an eclipse when the moon passes through the plane of the earth's orbit at new or full moon. If this happens, two weeks later it is still partially in the plane.

10. The earth's atmosphere scatters light from the photosphere. This scattered light hides the coronna.

11. The remaining part of the sun is still too bright, and sufficient light scatters in the sky to block our view of the corona. The photosphere is 1,000,000 time brighter than the corona, so even if it is 99% covered, we still have 10,000 times more light.

12. Primarily $H\alpha$ in emission, with some other lines like H, K and $H\beta$ mixed in.

13. The presence of emission lines caused by the stripping of more than a dozen of electrons off atoms; mostly x-ray radiation given off.

14. Ground-based: flexible, short lead times, state-of-the-art equipment, relatively inexpensive, can do many experiments simultaneously, can get finer resolution, can observe the current eclipses; satellite studies: can observe uv, x-rays.

15. An 11-year cycle of the number of sunspots on the sun. It can be explained by variations in the distribution of the solar magnetic field. The cycle is caused by the sun's differential rotation, with flux tubes being twisted and eventually kinking and pushing g of starlight; gravitational redshift; advance of perihelion of Mercury.

16. The solar wind consists of charged particles and is of very low density. It represents the outward expansion of the corona. The earth's wind consists of neutral atmospheric molecules and flows from place to place based on highs and lows.

17. (a) The solar constant is the amount of energy per second that would hit each square cm of the earth at its average distance from the sun, if the earth had no atmosphere. (b) Because of the varying transparency of the earth's atmosphere, the need to measure in other spectral regions, and the difficulty with accurate calibration of instruments. The instrument on Solar Maximum Mission has eliminated the first two of these problems. It has found variations in the solar "constant" by almost 0.2% within an interval of days.

18. visible.

19. Mars is 1.5 A.U. so receives $1/1.5^2 = 1/2$ the energy, so the solar constant is 1/2 that of the earth.

20. Bending of starlight, tested to better than 10% at eclipses and farther in the radio. Gravitational redshift, best verified in white dwarfs. Advance of perihelion of Mercury, matches observations; best verified for binary pulsar.

Chapter 22: Young and Middle-Aged Stars

1. By observing many stars, we see stars in various stages of their life cycles.

2. Gravitational collapse; when nuclear reactions start in the core.

3. The sequence of points on the H–R diagram that represents the star at various stages in its life.

4. Stays more-or-less steady on the main-sequence, with only a slight evolution.

5. Dark clouds; T Tauri stars; OB associations; pulsars.

6. $_1H^1$: 1p, 0n, 1e; $_3Li^6$: 3p, 3n, 3e; $_{26}Fe^{56}$: 26p, 30n, 26e.

7. (a) helium (another isotope); (b) hydrogen; (c) He III is already just a helium nucleus; there are no more electrons to move.

8. (a) High temperatures are necessary to overcome the electrostatic repulsion between the protons of the nuclei. (b) The proton-proton chain.

9. Chemical or gravitational energy. They don't provide enough energy for stars to last as long as they do.

10. Gravity pulling in vs. gas pressure pushing out.

11. $(1.495 \times 10^{11}$ m$)/(299{,}792$ m/sec $\div 60$ sec/min$) = 8$ min.

12. The average velocity of the particles in the gas is higher.

13. (a) $.007 \times 90\% = .006$. (b) $.006 \times 2 \times 10^{33} = 10^{31} = 2 \times 10^3$ earth masses (mass of the earth is in Appendix 2).

14. (a) Hotter central temperatures are needed for the carbon cycle than for the p-p chain. (b) They fuse hydrogen to helium at much greater rates than do less massive stars.

15. Energy of the positron, neutrino, and gamma ray.

16. Intermediate products of the first step are deuterium, a positron, and a gamma ray, of which the deuterium goes into making

He^3 plus a gamma ray. Two He^{3}s make one He^4 plus two H^{1}s. The net result is $4H^1 \rightarrow He^4 + 2e^+ + 2\nu + 2\gamma$.

17. The inner regions of the sun would start to contract. We would not know for millions of years, as it takes that long for photons to reach the surface, unless we were measuring the neutrino flow.

18. They are not stopped by the mass of the sun and reach us from the core at the speed of light.

19. They tell us that we may not understand the processes in stellar interiors or the behavior of neutrinos as well as we had thought we did.

20. It is sensitive to the majority of neutrinos emitted by the p-p chain.

Chapter 23: The Death of Stars Like the Sun

1. The end of hydrogen fusion in the center of the core.

2. All the hydrogen has been used up.

3. The core starts heating up, nuclear reactions start and generate enough energy to increase the pressure enough to support the star.

4. The helium flash briefly reverses the evolution to a red giant, which then resumes.

5. Because the electrons cannot be compressed any farther by the gravity available in that mass reange of a star.

6. $(1.5 \times 10^8 \div 0.7 \times 10^6)^2 = (2 \times 10^2)^2 = 4 \times 10^4$.

7. It occurs over a brief period in the evolution of a star.

8. The planetary nebula would appear slightly larger now.

9. We can see the hot core.

10. Pressure from electron degeneracy.

11. The sun is larger, has a cooler surface, and has nuclear reactions in its core.

12. About 0.5 solar masses. The rest will be ejected in a planetary nebula.

13. A white dwarf.

14. The hottest white dwarfs are similar in spectral type to an O star; the H-R diagram in the book shows some only up to A0. The

H-R diagram shows a white dwarf as cool as type K.

15. Show the sun going up to the right from the middle of the main sequence, then horizontally over to the left during the planetary-nebula stage, then at top left as a central star, and then down to the right to the white dwarfs.

16. We can measure spectral lines more precisely for the sun, but the effect is much greater for 40 Eridani B so is better measured there even though the mass is more uncertain.

17. In a nova, only a small amount of matter is fused, and the situation is not right for a chain reaction to continue.

18. IUE--advantages: can observe uv, can observe day and night; disadvantages: small telescopes, can't image, can't observe visible.

19. Evolution depends on mass, and so changes as mass is exchanged.

20. Matter in a star's Roche lobe is gravitationally bound to the star, but can flow through the neck to the adjacent star's Roche lobe and so change the mass of each and thus the evolution.

Chapter 24: Supernovae

1. They have so much surface area.

2. A nova is the ejection of a relatively small amount of matter from a star. A supernova is the explosion of an entire star.

3. It is relatively young. Otherwise it would have burned itself out.

4. It has stopped fusing hydrogen.

5. When iron fuses it takes up energy instead of releasing it. Thus, a supergiant's iron core provides no new energy to couteract the inward pull of gravity.

6. Most of it, perhaps 15 solar masses.

7. Each 5 magnitudes is a factor of 10^2. Twenty magnitudes is 4×5 magnitudes. $(10^2)^4 = 10^8$.

8. The spectra of Type II supernovae show many prominent hydrogen lines while spectra from Type I supernovae show no hydrogen. Type I supernovae brighten more than heavyweight supernovae.

9. In the best current models, a Type I is the incineration of a white dwarf in a binary system, and a Type II is the explosion of a massive star in runaway evolution.

10. From a H-R diagram, we can see that a B star is about 10 magni-

tudes or 10^4 times brighter than the sun. Thus the supernova brightens by $10^{11}/10^4 = 10^7$ times, which is brightening by 3×5 magnitudes for the 10^6 times and 2.5 magnitudes for the extra 10 times, making 17.5 magnitudes.

11. It contains a relatively large amount of nitrogen, and thus has processed material in the carbon cycle, in agreement with our theory for the evolution of Type II supernovae.

12. It will grow bigger, it will diffuse, and the filaments will change shape.

13. 2000 parsecs = 6600 light years. 1054 - 6000 = 5546 B.C.

14. High-energy nuclei travelling through space.

15. In the earth's atmosphere.

Chapter 25: Pulsars and Other Neutron Stars

1.
Object	Reason
regular star:	would see it
oscillating white dwarf:	observed period too rapid
oscillating neutron star:	observed period too slow
rotating white dwarf:	observed period too rapid, star would fly apart.

2. The Crab is rotating very rapidly, and since pulsars slow down with age, it must be one of the youngest. Also, since we believe it was formed in the supernova explosion of 1054 A.D., this association shows it to be that relatively young age. A different argument for rejuvenation applies to the millisecond pulsars.

3. The pulse's width is 0.05 sec, which means that if the object is oscillating, it is not larger than 0.05 light-second = 15,000 km. No limit is placed by the pulse width for a rotating object, though it would have to rotate once each second without being torn apart by centrifugal force. A more fundamental limit for a rotating object is that its surface cannot surpass the speed of light; $2\pi r < c \times 1$ sec; $r < 50,000$ km, smaller than the sun.

4. It had been thought that the youngest pulsars were spining fastest and were slowing down at the greatest rate; a new model (the best now involves rejuvenation) had to be found to explain the fast pulsar.

5. We are able to test the prediction of the general theory of relativity that the perihelion of an orbiting body will advance. The binary pulsar, whose perihelion advances 4°/yr, gives a much better test of this prediction than does Mercury, whose perihelion advances 43 arc sec/century. Also, the rate of change of the orbit of the binary pulsar indicates that the system is losing energy in the form of gravitational waves, the first indication of the existence of this prediction of the general

theory of relativity.

6. Einstein's general theory of relativity predicts that the presence of a mass affects the space around it, and the existence of gravitational waves is a consequence of that prediction.

7. Draw a figure-8 showing the small Roche lobe of the neutron star on one side and the larger Roche lobe of a massive companion on the other. Show gas going through the throat to fall onto the neutron star.

8. 4×10^4 km/s $\div$ 3×10^5 km/s = 0.1, so this shift is 0.1 or 10% of the speed of light. Thus the lines are shifted by 10%. The Hα line of hydrogen comes at 6563 A, so is shifted by about 660 A.

9. The jets of gas are being emitted by SS433 at great velocity, providing the main part of the Doppler shift, but the Doppler shift we see tells us only the radial velocity. Thus we measure a different velocity when we observe the jets at a different angle. As precession carries the jets around so that they are at differing angles from us, we thus see a varying Doppler shift.

10. We are seeing only the component along the radial direction (our line of sight), so the true velocity may be higher.

Chapter 26: Black Holes

1. The gravitational force is too strong.

2. A black piece of paper is black in the visible but may emit in other spectral regions.

3. (a) particle. (b) see Figure 21-31.

4. (a) lower; earth. (b) decrease; increase. In both red giants and white dwarfs the mass will remain relatively constant, and it will be the difference in distance from the center of mass that makes the largest effect on the strength of gravity.

5. (a) 30 km. (b) 10^{-2} cm.

6. Photon sphere: 18×4.5 km = 81 km in radius; event horizon: 18×3 km = 54 km.

7. The event horizon at maximum rotation is half that of a non-rotating black hole of the same mass, and is thus 40.5 km in radius. The stationary limit is 40.5 km in radius at the poles and 81 km in radius at the equator.

8. The photon sphere is 3/2 times the size of the event horizon. Escape from a rotating black hole: yes from the photon sphere and ergosphere; no from within the inner event horizon.

9. (a) No mass can escape but more can be added. (b) Through energy emitted in the form of elementary particles.

10. No. Black holes of high mass have low density at their event horizons.

11. X-rays might be given off from surrounding gas, but in the absence of a companion to allow us to measure the mass, we would be hard put to show that it was a black hole. On a larger scale, black holes in galaxies like that in M87 (described in Box 26.1) can be found by effects on other stars or by there being point-like brightness distributions.

12. When a single-line spectroscopic binary is present and indicates that the invisible companion contains more than about 5 solar masses.

Chapter 27: The Structure of Our Galaxy

1. We see arm-like structure in the distribution of young stars and from 21-cm observations. We note that other galaxies that have arms are spirals. Moreover, since other galaxies with similar gas and dust content are spirals, we must be one too.

2. The Milky Way would hardly be visible in the direction away from the center but would still be visible when looking toward the center.

3. Nuclear bulge, inner 2,000 pc radius; disk, next 10,000 pc out in each direction; halo extends at least as far out as disk and also far above and below the disk. Sun is 8 to 10 kpc out. See Fig. 27-3.

4. (a) Absorption: dust superimposed on a bright nebula or just on the general background of stars.
(b) Reflection: nebula that shines by starlight reflected off dust grains.
(c) Emission: gas is heated and glows.

5. If there is glowing gas, that makes an emission nebula, and if there is enough dust too, then can be a superimposed absorption nebula. The North America Nebula is an example.

6. The interior shape of North America is $H\beta$ and the absorbing part around it is dust.

7. Emission. The gas is glowing, giving off $H\alpha$ radiation.

8. We would see such stars if they were there.

9. X-rays are absorbed by even the thinnest parts of the atmosphere.

10. Sagittarius A.

11. There will be less infrared radiation generated within the telescope itself when it is cooled.

12. Infrared and radio astronomers detect a bright infrared and radio source at the galactic center only about 10 A.U. across. Infrared studies of the motion of nearby gas suggest that a giant black hole may be present that attracts the gas and makes it heat up and radiate in the infrared. VLA radio observations show a small bright spot at that location.

13. (a) H II regions, O and B stars (O-B associations), galactic clusters, gas and dust clouds.
(b) These are all "young" objects that are formed in the arms; we know from studies of other galaxies that H II regions are preferentially located in the spiral arms.

14. The elctron and the positron each have mass $\underline{m}$, so the total mass annihilated is $\underline{2m}$.

15. X-rays have sufficient energy per photon while infrared radiation does not.

16. Hot dusty objects; protostars.

17. Extremely hot objects, or objects like black holes in which matter will be accelerated to high velocities and thus high temperatures.

18. Discussion should include results of x-ray, gamma-ray, and infrared observations to tell us more about stars being formed, and high-energy processes in the galactic center and elsewhere.

19. We know very little, not even the direction from which they come. Only one burst has been pinpointed, and that one seems to have come from a supernova remnant in the Large Magellanic Cloud. If the others are in our galaxy and this is in the LMC, then it must have been atypically strong.

20. The compression of gas and dust from the density wave leads to gravitational collapse and star formation.

21. To see the spiral structure, we would probably have to be at least 1 galactic radius or 10 kpc out. That is over 30,000 l.y., and it would thus take 300,000 years to get there at $0.1\underline{c}$.

Chapter 28: The Interstellar Medium

1. (a) Radio observations allow you to "see" farther and through objects which are opaque in the visible. Can observe day and night. Not affected by weather. Can detect many molecules, whose spectral lines are primarily in this spectral region.
(b) Optical observations give you better angular resolution than single-dish radio observations; also, use of a photographic plate allows simultaneous observation of a whole region. Spectra of

stars can be studied.

2. About 10 times by number, 4 times by mass.

3. Spectral-line radiation does not follow Planck curves; 21-cm radiation is emitted from a transition between sublevels of cool hydrogen is an example.

4. The hot stars ionize gas to make an H II region. H I regions are regions of gas of higher density in between stars.

5. The redshift is a cloud's Doppler effect, and reflects its motion. The cloud's reddening is the effect by which redder wavelengths become relatively stronger than bluer ones in the continuum as well as in lines, and results from scattering off dust.

6. Depends on the presence of a hot continuum source behind the hydrogen cloud to get absorption; otherwise, we see emission.

7. Non-thermal; doesn't reflect a kinetic temperature (average particle velocity).

8. No. All points would be at rest with respect to each other, so no Doppler shifts would be observed.

9. The energy in the two states, before and after the spin-flip, is different. When the hydrogen atom spontaneously flips from the higher-energy state to the lower-energy state, energy is given off at 21-cm as an emission line. When 21-cm radiation is absorbed (an absorption line), the spin-flip can take place in the other direction. Deuterium has an analogous spin-flip line at a wavelength of 92 cm.

10. The transitions between interstellar hydrogen's lowest levels, which are the only levels that are excited at the low temperature of interstellar space, do not fall in the optical or radio parts of the spectrum.

11. OH. It was thought to be strange because the ratios of line strengths were not what was expected, which we now know resulted from maser action.

12. A relatively high population of electrons is built up on an upper energy level (a "population inversion") by pumping of some sort, and then a trigger makes these electrons jump down in energy simultaneously.

13. Astronomers were assuming that the atoms came together one at a time in space, which would make a low probability of multiple combination. Thus 3-atom combinations would be much rarer than OH. But other mechanisms, such as formation of molecules on the surfaces of dust grains, significantly increase the abundances of complicated molecules.

14. They provide a place to hold one or more atoms until others can stick to it to make complicated molecules.

15. CO.

16. The dust globules surrounding forming or just formed stars, as well as dust clouds from which stars may soon form.

17. The Orion Molecular Cloud is located behind the H II region that we know as the Orion Nebula. New stars are forming in the Orion Molecular Cloud, while the already-formed stars have ionized the gas that we call the Orion Nebula.

18. The Becklin-Neugebauer Object, one of the brightest infrared sources in the sky, is in the midst of the Orion Molecular Cloud. It is about 200 A.U. across and its temperature is 600 K. It is probably a very young star just barely operating on fusion.

19. IRAS found a ring of infrared emission in the Andromeda Galaxy, and brignt infrared emission from star formation in the Tarantula Nebula in the Large Magellanic Cloud. Molecular lines can also be observed in other galaxies.

20. Radio astronomers can observe all the time, 24 hours a day, since the sky does not scatter solar radio waves making it too bright to see through as it does for solar light.

Chapter 29: Galaxies

1. Elliptical.

2. Since viewing even a very elliptical object from the end makes a circular outline, the outline we see appears either as elliptical or less elliptical than the galaxy itself.

3. The SBb galaxy has its arms coming off a bar, whereas the Sb galaxy has its arms winding from the nucleus. Both sets of arms appear about the same.

4. Type E7 is now known to be the same as SO; the side view has a broad bulge; the bulge is more distinct at Sa, and diminishes through Sb and Sc.

5. For a rotating galaxy, the Doppler shifts of spectral lines from different regions of the galaxy will have different values. If the slit of the spectrograph is laid across the long axis of the galaxy, the spectral lines will appear tilted with respect to the dispersion.

6. Spiral galaxies have new stars forming. They are the only type with interstellar gas from which the stars can form. The direct evidence for star formation in spiral galaxies includes observations of such star formation in our own galaxy, IRAS discoveries of infrared sources in spiral galaxies and the Large Magellanic Cloud, and the discovery of interstellar molecules in the radio spectra of other galaxies.

7. Shape of tails seen in certain cases, multiple nuclei, existence of ring galaxies, huge faint shells of gas around some ellipticals.

8. Because of obscuration in our own galaxy, particularly in the plane of our galaxy. The nearby galaxies that are now being found are located near the galactic plane.

9. Observing supernovae, Cepheid variables, RR Lyrae stars, size of H II regions, link between a galaxy's infrared brightness and its speed of rotation, and apparent brightness of galaxies.

10. IRAS observations of spirals show radiation presumably from dust clouds heated by young stars; some galaxies are especially bright in the infrared.

11. The Einstein Observatory found that different clusters of galaxies have different distributions of x-ray emission. The x-rays come from a hot intergalactic gas. In some clusters, the gas is clumped, while in others it is spread out more smoothly and seems concentrated near the center. This effect may indicate age and evolution; individual galaxies may eject gas in younger clusters and this gas would spread out as the clusters age.

12. No. The Hubble law refers to galaxies, not their constituents.

Stars in our galaxy have peculiar velocities that are both positive and negative.

13. (for H_0 = 50 km/s/Mpc) 46 km/s;
approximately 3×10^6 ly = 1 Mpc, which corresponds to 50 km/s

14. (for H_0 = 50 km/s/Mpc) (a) 4s; (b) $v = H_0 d$; $d = v/H_0$ = 1000 km/s ÷ 50 km/s/Mpc = 20 Mpc.

15. $v = H_0 d$ = 50 km/s/Mpc × 0.1 Mpc = 5 km/s = 3 miles/s = 10,000 miles/hr.

16. $v = H_0 d$ = 50 km/s/Mpc × 10^6 ly × (1 pc/3.26 ly) × (1 Mpc/10^6 pc) = 50/3.26 km/s = 15 km/s. Alternatively, 10^6 ly = 1/3 Mpc, etc.

17. z = 0.2, therefore v = 0.2 c. $d = v/H_0$ = (0.2 × 3×10^5 km/s)/ 50 km/s/Mpc = $(0.6 \times 10^5)/(5 \times 10)$ Mpc = 10^3 Mpc = 1 Gpc.

18. "Radio galaxies" are "active galaxies" whose x-ray and radio emission is much greater than that of a normal galaxy like the Milky Way Galaxy.

19. Active galaxies often appear peculiar optically, such as having a jet like M87 or a distortion.

20. VLA, the Very Large Array, is a specific set of telescopes in Socorro, New Mexico; VLBI, Very Long Baseline Interferometry, is a technique used with various sets of telescopes distributed around the world. Both use the interferometry to get high-resolution results; the VLBI results are of higher resolution, but the VLA can make maps faster and can observe larger structures.

21. Because we are simultaneously receiving radiation from points separated by a greater distance, and it is the maximum separation that determines resolution. For a single-dish device, the maximum separation is the diameter of the dish or mirror, but for an interferometer, it is the baseline (to which the diameters of the dishes used add only a minor contribution).

22. VLBI: test of relativity (Section 19.10); nucleus of our galaxy (Section 24.2b, Figure 24-8C), nucleus of M87 (Box 26.1 and mention in Section 26.5b).

23. VLA: Cas A supernova remnant (Color Plate 52 and Figure 22-6); head-tail galaxy (Figure 26-27); double quasar (Figure 27-16); T Tauri (Figure 20-3); SS433 (Figure 22-24).

Chapter 30: Quasars

1. We need the position of the optical object in order to observe its redshift and to study its spectrum to find out such physical conditions as temperature and density.

2. (a) 6500-5500=1000 A, 1000/6500 × c = 50,000 km/sec. (b) $50,000.

ANSWERS TO QUESTIONS IN TEXT

3. For 0.1c: (a) 600 Mpc. (b) 23 cm.

4. about 10^5 km/sec (actually 9×10^4 km/sec).

5. For 3000 A - 4000 A: (a) 0.3c. (b) 2000 Mpc.

6. $(1 + 3.80)(1216$ A$) = 5837$ A; the text refers to a newer discovery at z=4.01: $(1 + 4.01)(1216$ A$) = 6092$ A.

7. $z = (1.85/0.15)^{1/2} = 3.5$; $\Delta\lambda = 3.5 \times 5000$ A $= 17,500$ A; $\lambda_{new} = 22,500$ A; emitted in visible (blue); observed in infrared.

8. $z = (1.95/0.05)^{1/2} = 6.2$; $\Delta\lambda = 6.2 \times 1216$ A $= 7,594$ A; $\lambda_{new} = 8,810$ A.

9. $0.2\underline{c} = 60,000$ km/sec; $d = v/H_0 = 1,200$ Mpc ($H_0 = 50$).

10. The extreme redshifts, visible in the optical spectrum.

11. It implies all that energy must be coming from a small region.

12. Pulsars are in our galaxy, are stars, pulse regularly, and are probably understood. Quasars are extragalactic, are much larger than stars, give off much more energy, pulse irregularly, and are not completely understood.

13. (a) If this is so, we should see some blueshifts, and we don't.
(b) This would require an unreasonably large mass, and we should detect other gravitational effects of such concentrations. Also, the gravitational effect should decrease with distance from the quasar for the "fuzz" detected around some quasars, and it has abeen found observationally that it doesn't.

14. If they are closer than we think, then they are not as luminous as we think. However, if the redshifts are accounted for by the Doppler effect, a large amount of energy is still necessary.

15. When one carefully considers the time interval over which light was actually emitted rather than the time interval between our observations, the speeds of the quasar components tuurn out to be less than $\underline{c}$.

16. In this chapter, we discuss only that dozens of quasars are x-ray sources, and that some of those are among the farthest known. We also point out that the x-ray observations are revising our understanding of the distribution of quasars. In the cosmology chapter, we discuss the contribution of quasars to the hitherto unresolved x-ray background; though the first thoughts had been that this x-ray background may come entirely from quasars, more recent work indicates that the quasars may make only a small contribution. The discussion bears on the missing mass problem and whether there is enough mass to close the universe.
    The new ability was the imaging x-ray telescope of the Einstein Observatory. We will have to wait at least until the 1990's for

the successor high-resolution x-ray telescope, AXAF.

17. Presence of structure near the bright quasi-stellar region, similar to a galaxy with a bright nucleus. There is a progression of galactic types from Seyfert and N galaxies, which appear more like quasars than normal galaxies.

Chapter 31:   Cosmology

1. It might have been noted even without observational evidence that an expanding or a finite universe would eliminate the paradox, but the chance of that happening seems slim (and indeed, it did not happen).

2. We would have to travel 10^{24} years. $10^{10}/10^{24} = 10^{-14} = 0.000\ 000\ 000\ 000\ 01\%$.

3. $E = hc/\lambda$; 1 mm/1 A = 10^7; the 1 mm photon has 10^7 times less energy.

4. The scale of clusters of galaxies; giant filaments and voids appear.

5. The expansion of the universe could be explained by either the big bang or the steady state theory.   (a) For the big bang theory: Background radiation.   (b) For the steady state theory: No hard evidence in favor.   Evidence against includes signs that the universe was different in different epochs, such as the existence of quasars and their distribution with redshift, the existence of the background radiation, and the explanation of the observed abundances of the light elements.

6. The big-bang models are solutions to the equations in Einstein's general theory of relativity; the theory explains how gravity affects space.

4. It assumes the universe is unchanging in time, a fundamental assumption some considered characteristic of "perfection."

5. If galaxies increase in luminosity as they age, they were fainter long ago.   When we look out, the far away galaxies appear fainter than they would be if their brightness did not change and thus are deduced to be farther than they actually are.   The rate at which we think the universe is expanding, on the assumption of constant brightness, is thus greater than the rate at which it actually is expanding.   The universe thus appears to be slowing less than it actually is.

9. $1/H_0 = [1 \div (50\ km/sec/Mpc)] \times [(3 \times 10^{19}\ km/Mpc) \div (3 \times 10^7\ sec/yr)] \times 10^6\ pc/Mpc = 2 \times 10^{10}$ yr.

10. The denominator is twice as large, so the answer is 1×10^{10} y. Taking gravity into account, the expansion would be slowed, and the value for the age of the universe would be larger.

11. The time when the universe became transparent.

12. $\lambda_{max} T = 0.3$ cm·K (Appendix 2), thus
    $\lambda_{max} = (0.3$ cm·K$)/(3$ K$) = 0.1$ cm $= 1$ mm.

13. About the motion of the sun and the Virgo Cluster in the frame of reference set by the overall material from which the last scattering of the background radiation came.

14. $\Delta\lambda = [(350$ km/sec$)\div(3 \times 10^5$ km/sec$)] \times (6563$ A$) = 7$ A.

Chapter 32: The Past and Future of the Universe

1. If the electron's charge is e, the positron's is −e; the electron and the positron have the same mass.

2. $m_e = 9 \times 10^{-28}$ g; mass annihilated is $2\,m_e$; so E $= (1.8 \times 10^{-29}$ g$) \times (3 \times 10^{10}$ cm/sec$)^2 = 1.6 \times 10^{-8}$ erg. Sun's luminosity is $(3.8 \times 10^{33}$ erg/sec$)\div(1.6 \times 10^{-8}) = 2 \times 10^{41}$ times greater, an enormous amount.

3. 1 hour $= 3,600$ sec; $_2\text{He}^4$: 0.1; $_1\text{H}^2$: 5×10^{-5}; $_1\text{H}^3$: 10^{-6}; $_4\text{Be}^7$ and $_3\text{Li}^7$: 10^{-10}; $_3\text{Li}^6$: 10^{-12}.

4. Deuterium is much more sensitive than helium to the density in the early universe; no deuterium has been formed since.

5. If the density in the early universe was great, the deuterium would have been quickly cooked into helium, and little deuterium would have been left today; in that case, there might be enough mass to close the universe. The observed deuterium abundance seems too high to allow a closed universe.

6. The deuterium abundance is too high to have a closed universe. The fact that a globular cluster and thus the universe is older than 16 billion years is not consistent with a closed universe for reasonable values of Hubble's constant. The idea that quasars might contribute all the x-ray background seemed to indicate that not enough matter can be hidden in the form of hot gas to close the universe, though Margon's restudy minimized or eliminated this contribution. On the other side, the evidence that neutrinos may have mass raises the possibility that a lot of matter is hidden in the form of neutrinos, thus closing the universe, though the original experiments showing neutrino oscillations have not, for the most part, been confirmed.

7. We would have to wait a long time between direct observations to detect a change in Hubble's constant.

8. An infinite amount, since the proton's mass would increase as its speed grew closer to the speed of light; $m = m_{rest}\div\sqrt{(1-v^2/c^2)}$. It thus takes more and more energy to accelerate the increasing mass.

9. The gravitational force is cumulative, and there is no anti-gravity (that is, no negative component).

10. It explains why the universe is so homogeneous, why we find no magnetic monopoles, why we may not be able to tell whether the universe is open or closed, and where most of the matter in the universe came from.

SAMPLE EXAMINATIONS WITH ANSWERS

This section includes 8 10-minute quizzes, 6 hour exams, and 3 final exams. Each of the quizzes and hour exams deals with part of the course. The last final exam is multiple-choice and fill-in.

10-Minute Quiz #1

1. Write in scientific notation: 380,000 km (the distance to the moon).

2. Write in scientific notation: Twenty-five million seven hundred thousand.

3. Write in scientific notation: .02.

4. Write in scientific notation: .00056.

5. Write out: 7×10^{-5}.

6. Write out: 6.54×10^3.

7. Write out: 6.54×10^{-3}.

10-Minute Quiz #2

1. Briefly describe 3 Voyager discoveries about Jupiter and its moons.

2. Briefly describe 3 Voyager discoveries about Saturn and its moons.

3. Briefly describe 3 Voyager discoveries about Uranus and its moons.

10-Minute Quiz #3

1. Sirius is magnitude -1.4. The faintest star observable at Palomar is magnitude 23.6. How many times brighter is Sirius than this faintest star? (Show your work.)

2. How many times brighter is a 1st magnitude star than a 7th magnitude star (approximately)? (Show your work.)

3. A difference of a factor of 125 in brightness is a difference of how many magnitudes?

10-Minute Quiz #4

1. The Hubble Space Telescope is now scheduled for launch in 1988. Since it will be above the earth's atmosphere, the sky above it will be very dark, and it will be able to detect stars of the 28th magnitude. At present, the darkest skies above a major observatory on earth are at the Cerro Tololo Observatory in Chile and the Mauna Kea Observatory in Hawaii, where one can see 24th magnitude stars.

How many times fainter are the faintest stars one should see with the Hubble Space Telescope?

2. How many times fainter are the faintest stars one should be able to see with the Hubble Space Telescope than the faintest stars one can see with the naked eye? Be sure to write down explicitly what the magnitude of these faint naked-eye stars is.

10-Minute Quiz #5

1. List the following in order of increasing wavelength:
   visible light, radio waves, x-rays, ultraviolet light, infrared.

2. Label each of the following spectra with its proper spectral class. One of the spectra is spectral class A and the other is spectral class C.

   : _____

   These strongest lines come from the element: _____

   : _____

   These strongest lines come from the element: _____

3. Which type of star is hotter: spectral class A or G?

10-Minute Quiz #6

1. Describe what IRAS was and discuss two of its discoveries.

2. What are the types of galaxies according to Hubble's classification?

3. What is an active galaxy?

10-Minute Quiz #7

1. How can globular clusters be used to locate the center of the Milky Way Galaxy?

2. Explain how both Cepheid variables and RR Lyrae variables are used to find the distances to globular clusters.

SAMPLE EXAMINATIONS WITH ANSWERS

10-Minute Quiz #8

1. Which is larger with respect to the wavelength of radiation that it is observing:

   A 10-cm diameter optical telescope used to observe light at a wavelength of 5000 A; or

   A 300-meter diameter radio telescope used to observe radio waves at a wavelength of 10 cm?

   Show your calculations. No credit will be given for an answer without any indication of how you got your answer.

2. Which has better resolution?

   The 300-meter diameter radio telescope used at a wavelength of 10 cm, or the same telescope used at a wavelength of 1 cm?

SAMPLE EXAMINATIONS WITH ANSWERS

Hour Exam #1

Do all questions. Partial credit will be given, so please show all
your work. Please write in pen, not pencil. You should be able
to go quickly through the whole exam. Do not dwell on any one
question. (Note: numerical values in parentheses are the point
values for that section of the question.)

1. a. The speed of light is 3×10^{10} cm/sec. What is the speed of
      light in km/sec? (3)
   b. During the Voyager explorations of Uranus, we had a graphic
      demonstration of the finite speed of light. Uranus is about
      20 A.U. from the Earth, and 1 A.U. = 150,000,000 km. How
      long after the radio signals left Voyager near Uranus did we
      receive them on Earth? (3)
   c. Write the diameter of the Earth, 12,756 km, in scientific
      notation. (2)
   d. How much is $10^6 + 10^7$? (Express answer in scientific
      notation.) (2)
   e. How much is $10^6 \times 10^7$? (Express answer in scientific
      notation.) (2)

2. a. Describe planetary rings in the solar system and how they
      are formed. How many are known and how were they
      discovered? (10)
   b. How did the discovery of Charon, Pluto's moon, lead to basic
      information about Pluto itself? Describe in detail. (10)

3. a. Discuss the role of the epicycle in the theories of Ptolemy
      and Copernicus. (10)
   b. Describe how Eratosthenes measured the diameter of the
      earth. (10)
   c. Describe and contrast the approaches of Kepler and Newton to
      the study of planetary orbits. (10)

4. --
              blue   red   (wavelength)

   a. The above line represents the electromagnetic spectrum.
      Above the line, label (in order) the visible, infrared,
      ultraviolet, radio, gamma-ray, and x-ray regions of the
      spectrum. (10)
   b. Does the left side of the line correspond to shorter or
      longer wavelengths? (1)
   c. Does the left side correspond to photons of higher or lower
      energy? (1)
   d. Does the left side correspond to higher or lower energy
      photons? (1)

5. (Note: All the answers to Question 5 should be in the form of
      numbers, and need only be approximate. All work must be
      shown, except possible conversions to scientific notation.)
   a. Venus can be as bright as magnitude -4. Betelgeuse is a
      first-magnitude star (m = +1). How many times brighter than
      Betelgeuse is Venus at its brightest? (4)

b. Star A has magnitude +12. Star B appears 10,000 times brighter. What is the magnitude of star B? (4)

c. Star C appears 10,000 times fainter than star A. What is the magnitude of star C? (4)

d. Star D has magnitude +10. The magnitude of star E is +8. How many times brighter does star E appear than star D? (4)

e. The Hubble Space Telescope will be able to detect stars of magnitude +28. The naked eye can see stars of about magnitude 6. How many times fainter are the stars that we will be able to observe with the Hubble Space Telescope? (4)

6. a. Describe and contrast how eclipses of the sun and of the moon occur. (6)

b. From which locations on earth can one see a total solar eclipse? The partial phases of a solar eclipse? A total lunar eclipse? (6)

SAMPLE EXAMINATIONS WITH ANSWERS

Hour Exam #2

1. a. Draw a graph showing a continuum with two emission lines. (5)
   b. Draw a graph showing two absorption lines. (5)
   c. Did a continuum have to be present or not in part b? Why? (3)
   d. Did a continuum have to be present or not in part a? Why? (3)

2. a. Explain how the same cloud of gas can be seen to cause an emission spectrum when seen from most angles, and can be seen to cause an absorption spectrum when seen from other angles. You may wish to draw a diagram to aid your explanation. (10)

   Are emission, absorption, or continuous spectra given off by:
   b. an ordinary incandescent light bulb? (1)
   c. a purple neon sign? (1)
   d. the solar photosphere (1)
   e. the red giant star Betelgeuse? (1)
   f. the moon? (1)

3. Which type of star has:
   a. the strongest hydrogen lines? (3)
   b. the strongest lines of ionized calcium? (3)
   c. lines of titanium oxide? (3)
   d. What does the nature of the spectrum of most stars (whether they are emission, absorption, etc.) tell you about how the temperature varies in those parts of the atmospheres of the stars where the spectrum is formed, and why? Be sure to specify what the nature of the spectrum of most stars is. (6)

4. a. What is the definition of parsec? (Note: A parsec works out to be 3.26 light years. This answer asks for the physical meaning of parsec, and not merely about its numerical value.) (5)
   b. What quantity describes whether a dying star becomes a white dwarf, a neutron star, or a black hole? (5)
   c. If you compare a photograph of a nearby planetary nebula taken 40 years ago with a recent one, how would you expect them to differ? Why? (5)

5. How can we determine the age of a galactic cluster by studying its Hertzsprung-Russell diagram? Draw an H-R diagram for a typical galactic cluster as part of your answer. Be certain to explain why the method you describe works. (20)

6. Why do we believe that pulsars are rotating neutron stars? Be complete and convincing. (20)

SAMPLE EXAMINATIONS WITH ANSWERS

Hour Exam #3

1. Identify briefly (3 points each):
   a. plage
   b. electron degeneracy
   c. dwarf star
   d. white dwarf star
   e. Ring Nebula
   f. negative hydrogen ion
   g. protostar
   h. solar wind
   i. SS433
   j. neutrino

2. a. Draw a cutaway diagram of the sun, showing all the significant layers inside and outside the surface. (5)
   b. List the layers, and give a characteristic temperature of each. (5)
   c. Sketch a graph of the sunspot cycle, labelling the axes and showing where we are now in the cycle. (5)
   d. Describe two ways in which the sun has helped verify Einstein's general theory of relativity. Which result was more important in view of the scientific method? Why? (10)

3. a. Make an H-R diagram and draw on it the future evolution of the sun. (8)
   b. What will the final state of the sun be? (4)
   c. Draw the equivalent of a person's evolutionary track on an H-R diagram for a person as the person ages, plotting height (increasing to the left on the horizontal axis) and weight (increasing upward on the vertical axis). Explain the various parts of the evolutionary track, labelling them with approximate ages. (5)
   d. On such a diagram as in (c), what would correspond to the main sequence of a regular H-R diagram for stars? Describe this, and sketch its location on the diagram. Be sure not to confuse a single evolutionary track with the whole main sequence. (3)
   e. Contrast the track you drew in (c) with the track of a star on an H-R diagram. (1)

4. a. Draw the main sequence of an H-R diagram, and, in addition to labelling its axes, label it with the masses of the stars. (6)
   b. Are massive main-sequence stars brighter or fainter than low-mass main-sequence stars? Why? (Your reason should be a fundamental one.) (4)
   c. Describe how the sun is now generating its internal energy. Write in detail. (10)
   d. What part of the sun does the neutrino experiment tell us about? (2)
   e. Describe the neutrino experiment and its results and interpretations. Comment on possible explanations for the results. (8)

SAMPLE EXAMINATIONS WITH ANSWERS

Hour Exam #4

1. a. What was the common property of all the Cepheid variable
      stars in the Magellanic Clouds that enabled Henrietta
      Leavitt to discover the period-luminosity relationship? (5)
   b. What are the Magellanic Clouds? Can they be seen from the
      United States? If not, why not? (5)
   c. Describe briefly what the period-luminosity relationship is,
      and what important property or properties it is ultimately
      used to derive. (10)

2. a. If you see a blue star surrounded by a blue nebula, would
      you expect this nebula to be emission or reflection?
      Explain. (10)
   b. If you see a blue star surrounded by a red nebula, would you
      expect this nebula to be emission or reflection? Explain.
      (10)
   c. Explain why some stars appear brighter on the blue plate of
      the Palomar Sky Survey (taken with the Schmidt telescope
      there) and other stars appear brighter on the red plate.
      What property of a star does this tell us? (10)

3. a. Sketch a side view of our galaxy as seen from outside it.
      Indicate the nucleus, the disk, and the position of the sun.
      (10)
   b. What kinds of objects are found in the galactic halo? Add
      the halo to the sketch you made in part (a). (5)
   c. What types of objects trace out spiral arms? Do these
      tracers follow the predictions of the density-wave theory?
      Why or why not? (10)

4. a. Estimate the wavelengths for which a window screen would be
      a good reflector. (5)
   b. List two advantages of radio astronomy over optical
      astronomy for the study of our galaxy. List two advantages
      of optical astronomy for the study of our galaxy. (10)
   c. What is "synchrotron radiation"? Why does it correspond to
      continuous radiation whereas transitions in a hydrogen atom
      give off spectral lines? (10)

SAMPLE EXAMINATIONS WITH ANSWERS

Hour Exam #5

1. a. Explain the density-wave theory of spiral structure and its advantage. (10)
   b. What is the difference between an H I region and an H II region? (5)
   c. Describe star formation in and near the Orion Nebula. (10)

2. a. How is a single-dish radio telescope used to make a map of one region of the sky? What is the difference between the procedures and our capability of observing with an optical telescope? (10)
   b. Describe the VLA. Give examples of results. (10)
   c. Describe VLBI. Give an example of a result. (5)

3. a. What is the spin-flip undergone by interstellar hydrogen? WHat radiation results, and how? (10)
   b. How is this radiation important to astronomers? (10)
   c. How is the way this radiation is formed in the hydrogen atom different from the way that recombination line radiation results? (5)

4. a. How can we detect an absorption nebula in space? (5)
   b. Why can't we see the center of our galaxy in visible light? (5)
   c. Describe the center of our galaxy. (10)
   d. What are two reasons for studying molecular lines? In what region of the spectrum do we best observe the radiation from these molecules? (5)

Hour Exam #6

1. a. Draw the Hubble classification scheme for galaxies, including labelled sketches of the different types of galaxies. (15)
   b. Sketch and describe what a cluster of galaxies looks like on the Palomar Sky Survey photographs. (10)

2. a. Describe progress in x-ray astronomy from the first observations of the sun and stars up to HEAO 1. (5)
   b. Describe the Einstein Observatory and its telescope. (5)
   c. Describe Einstein and radio observations of supernova remnants. (5)
   d. Describe Einstein observations of quasars. Discuss the implications of these observations. (5)

3. a. Describe the events that led up to the discovery of Hubble's law. (5)
   b. What is Hubble's law? What is Hubble's constant? (10)
   c. If a galaxy has its H alpha line shifted by 10% from 6563 A to 7119 A, how far away is it? (10)

4. a. Describe Olbers's Paradox and its resolution. (10)
   b. Describe evidence for the universe being open or closed. (15)

SAMPLE EXAMINATIONS WITH ANSWERS

Final Exam #1

Do all the identifications in Part I, and choose 6 out of 8 questions in Part II. Be succinct in your answers, but show enough of your thinking processes that we can give partial credit. Please write in ink.

I. Identify briefly: (3 points each)
   a. glitch   f. Great Red Spot
   b. dwarf   g. photosphere
   c. Olympus Mons   h. spicules
   d. Kitt Peak   i. diamond-ring effect
   e. light year   j. white dwarf

II. Choose six out of the following eight questions. (20 points each)

1. a. Draw a Hertzsprung-Russell diagram. Label the key parts with the names of the types of stars that correspond to those locations. Be sure to show the main sequence, and label the axes with the appropriate units. Mark the position of the sun.
   b. Show the evolution of a star of 1 solar mass from the time that it reaches the main sequence. Approximately how long does this 1 solar mass star stay on the main sequence?
   c. Describe the discoveries made by IRAS that have told us about star formation in our and other galaxies.

2. a. Draw a horizontal line and label it with the wavelengths of the electromagnetic spectrum (from left to right: 1 A, 10 A, 100 A, 1000 A, 1 micron, 10 microns, 100 microns, 1 mm, 1 cm, 10 cm and 1 m). Show on this line where the various named parts of the spectrum fall. (Mark the limits of each part of the spectrum.)
   b. Draw black-body (Planck) curves for gas at temperatures of 3500 K and 6000 K. Draw them to approximate scale on the same set of axes.
   c. How do Wien's displacement Law and the Stefan-Boltzmann Law show in the curves you have just drawn?

3. Give the approximate spectral types of stars of the following temperatures, and describe their spectra:
   a. 3000 K
   b. 6000 K
   c. 50,000 K
   d. Explain why we normally see absorption lines from the sun but see emission lines at an eclipse.

4. a. Describe the exit cone and its relation to the event horizon and the photon sphere of a black hole.
   b. Pulsars are best observed in what part of the spectrum?
   c. Describe the dispersion method as aplied to pulsars. What does it tell us?

5. a. Explain for each case how the Ptolemaic and Copernican models of the universe explain retrograde motion.
   b. State Kepler's laws.

c. Mercury's orbit has a semi-major axis of 0.4 A.U. How long does Mercury take to orbit the sun?

6. a. Describe three types of experiments carried on by the Viking missions to Mars.
   b. Discuss how the evidence from Viking bears on the question of life on Mars.
   c. When the Viking spacecraft was launched, it had to have a velocity of 40,000 km/hr to escape from the earth's gravity. It sent radio signals back to earth at a wavelength of approximately 10 cm; that is, they were at a wavelength of 10 cm when emitted by the spacecraft. At what wavelength were these signals received by antennas on earth? (Hint: the speed of light is 3×10^5 km/sec; you will have to convert km/sec to km/hr or vice versa before you apply the Doppler formula.) Be certain to show the Doppler formula.

7. List the planets:
   a. In order of increasing distance from the sun;
   b. In order of decreasing mass;
   c. In order of decreasing size. (Don't worry if you cannot fit in Pluto exactly or if two or more planets are similar in mass or size.)
   d. What is the origin of the lunar craters?
   e. Describe what we have learned about Uranus from space exploration.

8. Two stars in a binary system have orbital speeds of 10 and 20 km/sec. The period is 10 years. What are the masses of the two stars? (It is useful to know that 1 A.U. = 1.5×10^7 sec.) Hint: first find the radius of the orbits.

Final Exam #2

Choose 5 of the following 7 questions (20 points each). Please be short and sweet in your answers. Show enough of your thinking processes that we can give partial credit, but don't go on long enough to say something that is actually wrong. Please write in ink.

1. a. What is the cosmological principle?
   b. What is the perfect cosmological principle? How does it differ from the cosmological principle? How does it lead to the steady state theory?
   c. Why and how do we try to see to greater and greater distances in space in order to check whether the big bang or the steady state theory holds?

2. a. Discuss how quasars were discovered. The comparison of observations in what two parts of the spectrum led to their discovery? What breakthrough in analysis led to our current understanding of their properties?
   b. Why do some scientists argue that the quasars are not at the great distances that most people think they are?
   c. If they are at the closer distances, would that situation solve the energy problem? Why or why not?
   d. What is the possible connection between quasars and galaxies, and what new evidence indicates this link?

3. a. Identify VLA.
   b. Identify VLBI.
   c. Give one advantage of VLA over VLBI.
   d. Give one advantage of VLBI over VLA.
   e. What region of the spectrum do they both work in?
   f. Tell of at least one new observation that these techniques make possible.

4. a. What are two substantially different future courses the universe may take?
   b. Why do studies of interstellar deuterium give us information about this problem?
   c. Which future course can be predicted on the basis of the current deuterium observations?
   d. What is Olbers's paradox and what is its solution?

5. Consider a photographic plate of the spectrum of a distant galaxy. The plate has a dispersion of 20 Angstroms/mm.
   a. We measure the displacement--redshift--of a line whose rest wavelength is 6000 A to be 10 mm. How many Angstroms has the line been redshifted?
   b. At what speed is the galaxy moving with respect to us? Show your work, show the units you are using, show the value for the speed of light you are using, and make certain to show the units in which your answer is expressed. Make certain to write down explicitly the (non-relativistic) Doppler formula.

6. What kind of nebulae are
   a. The Ring Nebula?

  b. The Horsehead Nebula?
  c. The nebulosity in the Pleiades?
  d. Why were radio astronomers surprised a few years ago to start discovering all kinds of complex molecules in space, even in the tiny fractions of matter that they make up?
  e. How might these complex molecules form?

7. a. What type of galaxy do we live in? Why does the Milky Way appear as it does stretching across the sky? Illustrate with a sketch of the galaxy we live in as seen from outside, and illustrate the position of hte sun with an arrow.
    b. Describe how quasars were discovered and why they are thought to be unusual. What are two methods indicating that a point-like object in the sky is a quasar instead of an ordinary star?
    c. Describe the double quasar and its significance.

Final Exam #3
Multiple Choice and Fill-In

1. The back side of the moon is strikingly (a: similar to or different from) the front side. Tracking data from artifical satellites in orbit around the moon show (b: a uniform concentration of mass through the moon or mass concentrations under the surface). The moon is (c: like or unlike) the earth in having a crust differentiated from its core. The moon is relatively (d: less massive or more massive) than moons of other planets with respect to the mass of the planets.

2. Apollo 17 represented
   a. the beginning of a new lunar exploration program.
   b. the end of manned exploration of the moon for the present.
   c. the end of manned exploration of the moon for a 3-year gap.
   d. the end of all direct manned or unmanned lunar explorations.

3. The flow of heat from the lunar interior measured by astronauts on the moon is (a: higher or lower) than had been expected. This is one piece of evidence that the core of the moon is (b: solid or molten). The interior heat can come from (c: remnant heat from before the coalescence of the moon, radioactive elements in the interior, neither, or both).

4. The rocks that have been brought back from the moon give us a continuous picture from 4.6 billion years ago up to the present day (a: true or false). Collecting the lunar rocks was the primary reason that the space program was carried out (b: true or false). There are now permanent plans to set up an American space station on the lunar surface (c: true or false).

5. The moon can be seen (a: only at night or sometimes in the daytime or only with a telescope). A new moon occurs when the moon is (b: near the sun in the sky or far from the sun in the sky or half-way around the sky from the sun).

6. Two kinds of devices left on the moon by astronauts that are potentially of great value not only for astronomy but also for studies of the earth are ____(a)____ and ____(b)____ .

7. Kepler's third law shows that the period of revolution of a planet around the sun increases--that is, the planets take longer to go around--more (a: slowly or quickly) than the distances from the sun as we consider planet after planet. His law, in particluar, states that the ____(b)____ of the period of a planet is equal to the ____(c)____ of the distance. Uranus, which is sometimes barely visible to the naked eye at magnitude 5.7 in the evening sky, is about 20 A.U. from the sun; its period of revolution is thus about ____(d)____ . Is there a law relating the periods of revolution with the periods of rotation? If so, it is ____(e)____ .

8. The Mariner 9 spacecraft and the Viking orbiter photographed ____(a)____ and ____(b)____ on the surface of Mars besides sinuous rills. These sinuous paths seem to many scientists to indicate

the past presence of ___(c)___ on the Martian surface. Viking (d: has or has not) conclusively proved that life originated on Mars independently of life on Earth.

9. Planets known to have rings are: (list as many as there are).

10. To find the distance to the nearest stars we use their ___(a)___ . The distance to the nearest star clusters is found using ___(b)___ and ___(c)___ . To find the distance to the nearest clusters of galaxies we must assume that ___(d)___

11. Voyagers 1 and 2 travelled to the planets ___(a)___ and ___(b)___ They have safely passed the ___(c)___ belt. Voyager 2 went also to ___(d)___ .

12. Comets are made of ___(a)___ . They are visitors from other solar systems (b: true or false). Their tails can be (c: 100 km, 100,000 km, 100 million km, or 100 billion km) long. ___(d)___ Comet--the best-known bright comet--last visited us in ___(e)___ , and is due back in ___(f)___ . A comet shines because of nuclear fusion (g: true or false).

13. Telescopes with (a: wide or narrow) fields of view are best to search for new comets. The Palomar 5-meter (200-inch) telescope has a (b: wide or narrow) field of view. The Palomar 1.2-meter (48-inch) telescope has a (c: wide or narrow) field of view.

14. (a) Place in order of increasing wavelength: visible light, radio waves, ultraviolet, gamma rays, x-rays, and infrared. None of the light in the (b: name as many as applicable) reaches us on the surface of the earth because of the earth's atmosphere.

15. (a) Draw a reflecting telescope, showing the path of the light rays. (b) Why is the focusing element parabolic? Reflecting telescopes can be made bigger than ___(c)___ telescopes because ___(d)___ .

16. ___(a)___ is an example of a cluster of stars that shows the presence of gas and dust when you take a photograph through a telescope. Thus there is more material in space than just stars. ___(b)___ is an example of a supernova. This is a star (c: being born or dying).

17. Which telescope can provide better resolution when used by itself: the 91-meter (300-foot) radio telescope at Green Bank, or the 5-meter (200-inch) telescope at Palomar Mountain?

18. The sunspot cycle lasts about (a: 11 months, 11 years, 111 years). Other solar phenomena (b: do, do not) vary with that period.

19. The ___(a)___ source 3C 273 has recently been carefully studied. 273 is (b: its number in a catalogue, its right ascension, its declination, its celestial longitude, or its galactic coordi-

nates). It is a _____(c)_____ .

20. The sun is a _____(a)_____ -type star and has a surface temperature of
    _____(b)_____ K. In its next stage of evolution, it will be a _____(c)_____
    with a surface temperature of approximately _____(d)_____ . In its
    final state, it will be a _____(e)_____ of _____(f)_____ K. The sun's
    maximum brightness occurs in the _____(g)_____ stage.

21. Draw a graph of a spectrum with: (a) an emission line, (b) an
    absorption line, and (c) the continuum.

22. In the largest telescopes, such as the 3-meter (120-inch) tele-
    scope at the Lick Observatory, under the highest magnifications a
    quasar looks like (a faint planet, almost like a star, a dwarf
    galaxy, a spiral galaxy, or a planetary nebula).

23. Cygnus X-1 is thought to be a _____(a)_____ because when we study it
    in detail, we find that there is an invisible object present whose
    _____(b)_____ is more than _____(c)_____ times that of the sun. Also,
    radiation observed in the _____(d)_____ part of the spectrum
    fluctuates rapidly.

24. The IRAS satellite observed _____(a)_____ radiation and is best
    known for its observations of the _____(b)_____ part of the spectrum.
    The Einstein Observatory observed in the _____(c)_____ part of the
    spectrum. The Space Telescope will observe in the _____(d)_____ and
    _____(e)_____ parts of the spectrum.

25. Superluminal velocities have been discovered in (a: pulsars,
    quasars, or black holes). The objects for which superluminal
    velocities have been found move (b: as fast as, not as fast as,
    faster than) the speed of light. The phenomenon (c: does, does
    not) violate _____(d)_____'s special theory of relativity.

26. Hubble's law relates _____(a)_____ and _____(b)_____ . Hubble
    discovered it by measuring these quantities for dozens of quasars
    (c: true or false). (d: Maarten Schmidt, Halton [Chip] Arp,) is
    currently championing the validity of Hubble's law, while the
    other is opposing it. The Hubble relation (e: proves or does not
    prove) that the universe has a center.

27. List in order of increasing distance from us:
    quasars, pulsars, Barnard's star, the center of our galaxy, and
    the Andromeda galaxy.

28. Most evidence suggests that quasars are not really at the
    distances derived for them using Hubble's law (a: true or false).
    We have to measure the spectrum of each star on the photographic
    plate to tell which ones are likely to be quasars (b: true or
    false).

29. Novae occur in (a: newly born, recently exploded, newly
    brightened) stars. Specifically, the nova can happen to a star in
    the _____(b)_____ stage of its life, and only if _____(c)_____ is
    present.

30. The spiral arms of our galaxy are traced out by _____(a)_____,
    ___(b)___ , and ___(c)___ . These tracers indicate
    that stars are (d: being born, dying, in their red-giant stages)
    in the arms.

31. The ___(a)___ (b: atom or molecule) has a spectral line at a
    wavelength of 21 cm. We can see (c: further or not as far) in
    our galaxy in 21 cm than in visible light. This line is called
    Lyman alpha (d: true or false). ___(e)___ , heavy hydrogen, has
    a spectral line at a wavelength of 92 cm. It is a (f: higher or
    lower) frequency line than 21 cm.

32. The Pleiades form an example of a(n) ___(a)___ cluster. It
    contains relatively (b: young, old) stars, surrounded by
    (c: emission, reflection) nebulae. The stars contain relatively
    (d: small, large) amounts of heavy elements.

33. Many molecules have recently been detected in interstellar space
    besides molecular hydrogen and OH, such as ___(a)___ ,
    ___(b)___ , and ___(c)___ . It has proved more fruitful to work
    at (d: longer or shorter) wavelengths to discover new molecular
    lines. This requires (e: larger or more accurate) telescopes.
    The molecular measurements are useful because they allow us to
    tell the ___(f)___ and ___(g)___ in interstellar clouds.

34. The Orion Nebula is a(n) ___(a)___ nebula, also known as a(n)
    ___(b)___ . The North America Nebula is a(n) ___(c)___
    and a(n) _____(d)_____ at the same time.

35. Draw a hydrogen atom. Draw a deuterium atom. Use an open circle
    for a neutron, a solid circle for a proton, and a - sign for an
    electron.

36. The main components of the Milky Way Galaxy are (list four in
    order of increasing distance from the galactic center).

37. List in order of the openness of the spiral arms from relatively
    closed to relatively open:
    Sa, Sb, Sc, elliptical.

38. Wilson and Penzias received the Nobel Prize in Physics for their
    discovery of ___(a)___ . Their discovery is thought to be a
    remnant of ___(b)___ , and endorses the ___(c)___
    theory.

39. An absorption line (a: can, cannot) occur without a continuum. An
    emission line (b: does, does not) need a continuum to be seen. At
    a total solar eclipse the sun shows (c: a continuum, emission
    lines, or absorption lines).

40. Place in decreasing order of their masses:
    white dwarfs, black holes, neutron stars, planets, and
    galaxies.

41. Place in decreasing order of their average densities:
    white dwarfs, black holes, neutron stars, planets, and galaxies.

42. White dwarfs are (a: hotter or cooler) and (b: larger or smaller) than the sun.

43. Most pulsars are being studied in the ___(a)___ region of the spectrum. They are now being identified with ___(b)___ stars. A typical pulsar is ___(c)___ .

44. The pulsar in the Crab Nebula is unusual because it pulses not only in the ___(a)___ part of the spectrum, but also in the ___(b)___ and ___(c)___ regions of the spectrum. The fastest pulsar pulses about every ___(d)___ . Pulsars may be remnants of ___(e)___ .

45. Infrared light is intercepted largely by ___(a)___ in the earth's atmosphere. Stars that are (b: hotter or cooler) than average have their radiation peaks in this region. ___(c)___ is a particularly strong IR source in our galaxy.

46. Gamma rays are (a: longer or shorter) wavelength radiation than x-rays. The new satellites being launched to study x-rays and gamma rays are called ___(b)___ and ___(c)___ .

47. SS433 is an odd x-ray binary. We think it is a ___(a)___ surrounded by a ___(b)___ . Its apparent rapid movement away from us and toward us at the same time is the result of ___(c)___ .

48. The VLA is used to observe in the ___(a)___ part of the spectrum. It gives astronomers observations with high ___(b)___ .

49. When you are shining your flashlight back to earth as you travel into a black hole, at the point where the light will no longer escape you have passed the ___(a)___ We think that Cygnus X-1 is probably a black hole because its ___(b)___ is too (c: high or low) for it to be a neutron star.

50. The next eclipse of the sun will reveal the ___(a)___ and ___(b)___ of the sun that are not normally visible. The phase of the moon will be ___(c)___ . A solar eclipse occurs somewhere in the world every (d: month, year, decade) or so.

ANSWERS TO SAMPLE EXAMINATIONS

10-Minute Quiz #1

1. 3.8×10^5

2. 2.57×10^7

3. 2×10^{-2}

4. 5.6×10^{-4}

5. 0.00007

6. 6,540

7. 0.00654

10-Minute Quiz #2

1. Extent of magnetic field, wind speeds in belts and zones, rotation of red spot, a ring, volcanoes on Io, surface characteristics of other moons, etc.

2. 100,000 ringlets, material in apparent "gaps" in rings, role of shepherd satellites, wind speeds in belts and zones, nitrogen atmosphere of Titan, surface characteristics of other moons, etc.

3. Offset and wobble of magnetic field, rotation period from clouds, latitude effect in clouds, equal temperature at N and S poles, two new rings, strange surface of Miranda, cratering on moons, etc.

10-Minute Quiz #3

1. $-1.4 - 23.6 = 25$ magnitudes. Each 5 magnitudes = 100 times; therefore, 25 magnitudes = $(10^2)^5 = 10^{10}$ times.

2. Difference of 6 magnitudes = $2.5 \times 100 = 250$ times.

3. Factor of 125 is factor of 100 times 1.25. Factor of 100 is 5 magnitudes. Factor of 1.25 is less than factor of 2.5 (= 1 magnitude), and is about a factor of $\sqrt{2.5}$ (= 0.5 magnitudes) = 1.5, thus about 1.3 magnitudes.

10-Minute Quiz #4

1. With the ST we can see 4 magnitudes = 40 times fainter.

2. Naked-eye limit is about 6th magnitude. $29 - 6 = 23$ magnitude difference = $(2.5)^3 \times (10^2)^4 = 15 \times 10^8 = 1.5 \times 10^9$.

10-Minute Quiz #5

1. X-rays, ultraviolet light, visible light, infrared, radio waves.

2. Top spectrum: A star, hydrogen lines strongest. Bottom spectrum: G

star, (ionized) calcium lines strongest.

3. The A star is hotter.

10-Minute Quiz #6

1. IRAS was the international Infrared Astronomical Satellite that mapped the infrared sky during 1983. Its discoveries included comets, asteroids, galaxies, radiation from globules (indicating star formation), and signs of star formation in the Large Magellanic Cloud and the Andromeda Galaxy.

2. Hubble-Sandage: E0-(E7=S0); Sa-Sc; SBa-SBc.

3. Galaxies that radiate much stronger in the radio and/or x-ray parts of the spectrum than "normal" galaxies.

10-Minute Quiz #7

1. The clusters form a halo-shape that is centered on a point quite distant from the sun; it is reasonable to assume that that point must be the center of our galaxy.

2. From the Cepheid period-brightness relationship, or from the approximate same absolute magnitude of all RR Lyrae variables, we deduce the absolute magnitude of at least one star in the cluster. We can easily measure the apparent magnitude by observing, and from the difference can calculate the distance to the star and thus to the cluster.

10-Minute Quiz #8

1. $10\text{cm} / 5000 \text{ A} \times 10^8\text{A} / \text{cm} = 10^9 \text{ A} / 5 \times 10^3 \text{ A} = 2 \times 10^5$ times.

2. The telescope used at the shorter wavelength; that is, at 1 cm.

ANSWERS TO SAMPLE EXAMINATIONS

Hour Exam #1

1. a. 3×10^{10} cm/sec × (1 km/10^5 cm) = 3×10^5 km/sec.
   b. 20 A.U. × 1.5×10^8 km/A.U. = 3×10^9 km one way at the speed of light (part a), so the trip takes about 10^4 sec = 3 hours.
   c. 12,742 km = 1.2742×10^4 km.
   d. $10^6 + 10^7 = 0.1 \times 10^7 + 10^7 = 1.1 \times 10^7$; or $10^6 + 10 \times 10^6 = 11 \times 10^6 = 1.1 \times 10^7$.
   e. $10^6 \times 10^7 = 10^{6+7} = 10^{13}$.

2. a. Rings are known to exist around Saturn, Uranus, and Jupiter. They are formed by tidal forces, when matter is inside a planet's Roche('s) limit. Saturn's rings were discovered by Huygens some years after Galileo had noted that Saturn was not round (had "ears"); the Voyagers discovered thousands of ringlets. The 9 rings of Uranus were discovered during a stellar occultation. The ring around Jupiter was discovered on a Voyager 1 photograph. Additional details might include the presence of spokes and shepherding satellites in Saturn's rings and the temporary braiding of Saturn's F-ring. A partial ring or partial arcs apparently exist around Neptune.
   b. From the separation of Charon from Pluto and the period of revolution we can calculate the mass (and thus the density) of Pluto itself.

3. a. Both and Copernicus had epicycles, though Copernicus was able to dispense with the equant. Copernicus needed the epicycles because he was still using circular orbits; only with the work of Kepler did it become known that the orbits are elliptical.
   b. At noon on the day of a solstice, he measured the angle of the projection of a gnomon by measuring the shadow. Since he knew how far north this location was of another location where the sun was exactly overhead (no shadow), he could tell how much curvature the earth had. Since he knew the distance paced off between the locations, he could measure the size of the earth.
   c. Kepler was empirical, searching for regularities in the data. Newton took a broad theoretical aproach, and derived Kepler's laws as consequences of his more general theory of gravitation.

4. a. Gamma-ray, x-ray, ultraviolet, visible, infrared, radio.
   b. Shorter.
   c. Higher.
   d. Higher.

5. a. Mag +1 to mag −4 is 5 mag = 100 times.
   b. 10,000 times = 10^4 times = 10 magnitudes, so star B has magnitude +2.
   c. 10 magnitudes fainter than mag +12 is mag +22.
   d. 2 magnitudes difference is about $2.5 \times 2.5 = 6$ times.
   e. 28 − 6 = 22 magnitudes, which is 4 × 5 magnitudes + 2 magnitudes = $(10^2)^4 \times 6.2 = 6 \times 10^8$ times.

6. a. An eclipse of the sun is the projection of the shadow of the moon onto the earth. Since the moon's shadow is only about 200

km wide when it hits the earth, the band of totality is narrow and long (since it sweeps across the earth). An eclipse of the moon occurs when the moon is physically in the earth's shadow, so wherever on earth the moon is visible the lunar eclipse will be seen.

b. The path of totality is 200 km or so wide and thousands of kilometers long. The partial phases are visible for thousands of miles to either side of the band of totality. The total lunar eclipse is visible wherever the moon is up and not obscured by clouds at the time of the eclipse.

Hour Exam #2

1. a.   b.

c. Yes, because absorption lines have to absorb into something, i.e. the continuum.

d. No, emission lines can well exist by themselves.

2. a. Gas in a cloud under consideration gives off emission lines when its atoms change from higher energy levels to lower ones. But when we look through the cloud at a source of continuous radiation that is hotter than our cloud, our cloud will absorb radiation at certain wavelengths as the atoms in our cloud jump to higher energy states. Thus we might see absorption spectra if we look from point A, but emission spectra if we looked from other points such as B and C.

b. Continuous.

c. Emission.

d. Absorption.

e. Absorption.

f. Absorption.

3. a. A is acceptable (historically the spectra with the strongest hydrogen lines got assigned an A). The peak actually occurs between B9 and A0.

b. G is acceptable, since the sun's H and K lines are strong and it is a G star. K stars have equally srong H and K lines, though.

c. M (molecules can exist only in the coolest stars).

d. Most stars have absorption spectra, which shows, by the considerations in question 2, that continuous radiation is passing through a cooler gas. Thus the temperature of the stellar gas decreases as you go outward from the parts of the star where the spectrum is formed. (Note that the continuum is formed in the lower part of the photosphere, and not in the interior, as students sometimes answer.)

4. a. A parsec is the distance at which the angle between the earth and sun seen at maximum separation appears to be 1 second of arc.
   b. Mass.
   c. The planetary nebula is expanding rapidly, and will appear slightly bigger.

5. Since we can tell how long stars of given spectral types (and thus masses) will live on the main sequence, we can look at the H-R diagram for a galactic cluster and see how old the stars are at the place where the data begin to turn off the main sequence. Thus we know the age of the cluster, as we can reasonably assume that all the stars in the cluster were formed at approximately the same time.
   The drawing should have labelled axes and should show a main sequence where stars turn off to the upper right.

6. The determination that pulsars are rotating neutron stars depends on the process of elimination, and thus must be logically complete. First, all the possibilities must be enumerated, with some kind of statement that we have a complete list. Then each possibility must be dispensed with:
   Double stars ruled out because they could not revolve around each other so fast unless they were very compact, in which case they would emit enought gravitational radiation to cause observable changes in the periods of their revolution (not so much detail necessary for credit);
   Pulsation ruled out because the pulsation period depends on density, and white dwarfs and ordinary stars would pulsate too slowly while neutron stars would pulsate too fast;
   Rotating white dwarf ruled out because it would be torn apart rotating every 0.25 second, much less than the 0.33 seconds that the Crab pulsar takes to pulse.
   Finding the pulsar in the center of supernova remnants (Crab pulsar, Vela pulsar) is a clincher, but not the whole reason. Having the energy to illuminate the Crab Nebula match the abount that would be lost by a rotating neutron star in the Nebula is also a clincher but not the logical reason.
   (Note: a mere description of the lighthouse model is not a reason why we believe that model.)

Hour Exam #3

1. a. A bright area on the solar surface seen in Hα.
   b. The quantum-mechanical force that prevents the electrons from being compacted beyond a certain point, creating a pressure to

counterbalance gravity.

c. A normal main-sequnce star.

d. A small (size of earth), dense (tons per cubic centimeter), under-luminous star, located to the lower left of main sequence on H-R diagram.

e. A **planetary** nebula; a shell of glowing gas surrounding the star that blew it off. The Ring Nebula is in the constellation Lyra.

f. H^-, one proton with two electrons.

g. A primitive star, a collapsing ball of gas and dust that will ultimately begin nuclear fusion and thus become a star.

h. The expansion of the solar corona into interplanetary space; ions and electrons flowing out from the corona, probably from coronal holes.

i. A neutron star emitting jets of gas at 25% the speed of light.

j. A spinning particle with no charge that has no rest mass (unless recent experiments that may indicate a neutrino mass are correct) and (if it is massless) always travels at the speed of light; it is emitted in nuclear reactions. Neutrinos are elusive (hard to capture), but that is not a defining property.

2. a and b. Core: 15×10^6 K. Photosphere: 6000 K. Chromosphere (and spicules): 15,000 K (ok: a temperature as low as 7500 K). Corona: 2×10^6 K.

c. The graph should have time in years on the horizontal axis and sunspot number or some measure of solar activity on the horizontal axis. It should show an oscillation with an 11-year period.

d. The annual advance of Mercury's nearest point to the sun and the deflection of light from stars near the sun during a total solar eclipse. Since the second effect was predicted by Einstein's theory without having been previously observed, it was a more significant confirmation of the theory in view of the scientific method. A third effect is the gravitational redshift of light from the sun, but this effect is barely noticeable. It would be more important than the advance of Mercury's perihelion but less conclusive than the eclipse experiment.

3. a. The sun should move first to the upper right, then horizontally across to the left, and then down to become a white dwarf.

b. White dwarf.

c. On the plot, a baby will gain height and weight and thus will move from lower right to upper left, and then reach a certain point in the teenage years and stay relatively constant there through maturity. Some people put on weight as they go through middle age. Possible nuances include dieting, or the loss of some weight or height for the elderly.

d. The main sequence would be a band extending from lower right to upper left, corresponding to the general observation that taller people weigh more than shorter people. The main sequence must not be confused with the single evolutionary track drawn in (c). That single track should hit the main sequence at one point and stay there approximately throughout maturity.

e. A person's evolutionary track is fairly monotonic, with rapid height and weight increase at first and some modifications toward the end of the person's life. A star's pre- and post-main sequence tracks are much more complicated. A similarity is the long period that a star and a person remain on each main sequence.

4. a. At upper left (O stars), up to 30-50 solar masses. Sun is 1 solar mass. M stars at extreme lower right would be 0.1 solar masses.

b. Brighter, because the larger mass has led to greater energy release in the gravitational contraction, with corresponding higher temperature and much more energy generation. Additional point: higher energy generation is necessary to balance stronger gravity; CNO cycle is in use.

c. The sun is now generating its internal energy through the proton-proton chain, in which four hydrogen atoms combine sequentially to form one helium atom (actually six hydrogen atoms form one helium plus two hydrogens left over), with 0.007 of the mass transformed into energy according to $E = mc^2$.

d. The core.

e. The neutrinos rarely interact with matter, so a large tank of 400,000 liters of C_2Cl_4 is located deep undergroundand thus can speed their way out of the core much more rapidly than photons of energy, which take millions of years and millions of absorptions and emissions by atoms to get to the solar surface. But a new theory shows that 2/3 of the neutrinos emitted could change type en route to earth, and thus not be picked up by our detector. New types of detectors should improve this situation.

ANSWERS TO SAMPLE EXAMINATIONS

Hour Exam #4

1. a. Their distance. (Thus the relation of the apparent magnitudes of the stars with respect to each other is the same as the relation of the absolute magnitude.)
   b. Two galaxies. (Satellite galaxies of the Milky Way; the nearest galaxies to us.) They cannot be seen from the U.S.; they are too far south in the sky. (Their declinations are too low.)
   c. For Cepheid variables, the absolute magnitudes, that is, the intrinsic luminosities of the stars, can be told by observing the periods of the variation. There is a graph on which a simple curve, almost a straight line, links the period and the absolute magnitude. From the period-luminosity relation, we can derive the distance.

2. a. Reflection. An emission nebula would show strong red (Hα).
   b. Emission. We can see the hydrogen (Hα) radiation.
   c. The hotter stars have the peak of their radiation in the ultraviolet and are thus brighter on the blue plate. Cooler stars peak in the red, and thus appear brighter on the red plate. From comparison of the plates, we can tell the temperature of the stars.

3. a.

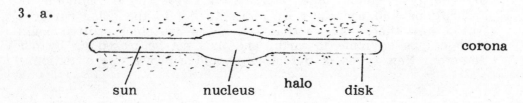

   b. Globular clusters.
   c. H II regions, O and B stars, galactic clusters. The density-wave theory predicts that stars should form as the density wave passes by, so we expect to see hot, young stars.

4. a. The mesh of a window screen is, offhand, about 1/4 cm, so reflects radio waves longer than about 1 or 2 cm. (In English units, 1/8-inch mesh, thus waves longer than about 1 inch.)
   b. Advantages of radio astronomy over optical astronomy: can see through to galactic center and to other side of galaxy; can observe day and night; not troubled by weather; VLBI techniques give us higher resolution; can study active galaxies.
   Advantages of optical astronomy over radio astronomy: can measure spectral types and study surfaces of stars; have better resolution than single-dish radio astronomy; can study variable stars like Cepheids; can see stars from light emitted at their radiation peaks; basically see different types of objects.
   c. Synchrotron radiation is continuous radiation given off as electrons spiral around magnetic field lines at relativisitic speeds. The energy the electron can have is not quantized, and hence there is no set of energy levlels between which the electron must jump, unlike the case for the enrgy levels in a hydrogen atom. As E=hν, since the energy is not quantized, the result of energy changes is not quantized into discrete wavelengths known as spectral lines.

ANSWERS TO SAMPLE EXAMINATIONS

Hour Exam #5

1. a. We observe a differential rotation of stars and gas that would
      have caused spiral arms of galaxies to wind up many times,
      contrary to observation. Hence we postulate that a density wave
      of spiral form circles the galaxy like a pinwheel, and that the
      spiral arms that we see merely show us where the density wave
      has passed recenly. When the density wave passes, it compresses
      the gas and dust, and new stars form.
   b. The hydrogen is neutral in an H I region, and ionized in an H II
      region.
   c. The Orion Nebula is an H II region, behind which is a molecular
      cloud in which star formation is going on. The BN object is a
      young hot star, and other infrared objects and objects we detect
      as radio water masers are also present.

2. a. At a given frequency, we point to a location in the sky and
      measure the strength of the radiation from there. To build up a
      map, we must measure point after point. With an optical
      telescope, we could use a photographic plate to make an image
      all at once. (Some may compare this radio mapping with
      photoelectric rather than photographic optical methods.)
   b. An array of 27 radio telescopes stretching in a "Y" 27-km in
      diameter across a plain in New Mexico. Makes images of objects
      like supernova remnants and planetarey nebulae with resolution
      comparable to the best ground-based resolution.
   c. Interferometry carried out with two or more radio telescopes
      stretched great distances, even continents, apart. Resolution
      1000 times better than best optical; can study point source at
      center of our galaxy, in M87, and superluminal expansion of
      quasars.

3. a. The direction of the spins of the electron and proton in a
      hydrogen atom change from parallel to antiparallel or vice
      versa. Since energy is given off or taken up when this happens,
      a spectral line is formed. Its wavelength is 21 cm (1420 MHz).
   b. We use it, together with the differential rotation curve of the
      Milky Way Galaxy, to map out the spiral arms in our own galaxy.
   c. The spin-flip causes a splitting in the lowest energy level of a
      hydrogen atom. Recombination lines result when an electron that
      has been separated from its proton rejoins a free proton,
      starting on a very high energy level and then jumping down to
      lower energy levels.

4. a. By seeing how far it dims the light from stars behind it; more
      particularly, but noticing that there seem to be fewer stars
      than average in that region of the sky.
   b. We can't see through all the dust.
   c. A high-mass, very small source giving off strong radio and
      infrared radiation is concentrated there; perhaps it is a giant
      black hole.
   d. They tell us the conditions in dark clouds; they tell us that
      conplicated molecules can form easily lin space (and thus that
      it was not difficult for the precursors of life to form). We
      study them in the radio, for the most part. Molecular hydrogen
      is seen in the ultraviolet.

ANSWERS TO SAMPLE EXAMINATIONS

Hour Exam #6

1. a.

b. (This question is based on the lab given earlier in this book.) The sketch should show a variety of types of galaxies, and a variety of orientations.

2. a. Giacconi's rockets discoverd Scorpius X-1; rocks and Skylab solar x-ray observations in the early 1970's discovered coronal holes; Uhuru mapped hundreds of sources; x-ray sensitivity was too low to detect many stars until HEAO-1, which expanded the number of objects to about 1500. Einstein Observatory (HEAO-2) and EXOSAT studied individual objects.

b. A grazing incidence telescope with resolution of about 2 arc sec, feeding several x-ray instruments, including an imager.

c. Einstein x-ray and VLA radio observations of supernova remnants like Cas A could be compared; the shock front where the supernova met the interstellar medium could be seen. X-ray spectra show enhanced abundances from nucleosynthesis. Einstein observations of the Crab show the central pulsar.

d. Einstein discovered many serendipitous x-ray sources in otherwise blank fields; many of the sources were new quasars. It had been hoped that they could account for the x-ray background; they seem to provide a substantial part but perhaps not all.

3. a. Slipher discovered that galaxies had large redshifts. Hubble then measured many more redshifts and associated distances, and realized that they were correlated.

b. The velocity of recession is proportional to the distance; Hubble's constant is the constant of proportionality: $v = H_0 d$.

c. It is receding at $0.1c = 30,000$ km/s. $d = (30,000$ km/s$)/(50$ km/s/Mpc$) = 600$ Mpc.

4. a. Olbers' (not Olber's) Paradox: since in an infinite universe we should eventually see a star no matter in which direction we look, why are there dark spaces between the stars in the night sky? Resolutions: (1) The expansion of the universe leads to a diminution of the energy in each photon, and (2) before we should eventually see a star, we have to see out so far that we are seeing back before the stars formed.

b. Open: the deuterium abundance is too high for the universe to be closed; if the density was high enough for the universe to be closed, then the deuterium would have been transformed into helium in the early minutes after the big bang. The motion of nearby galaxies shows that there is not enough mass in the Virgo Cluster between the visible mass for the universe to be closed. The evidence for an open or closed universe from study of dsitant galaxies in the Hubble diagram depends too much on evolutionary considerations to be reliable.

Final Exam #1

Part I

1. a. A jump in the period of a pulsar.
   b. A normal main-sequence star.
   c. The largest volcano on Mars.
   d. The site of the U.S. national optical observatory.
   e. The distance that light can travel in one year.
   f. A long-lasting red spot on Jupiter; probably a giant hurricane.
   g. The layer of the sun from which the visible light is emitted.
   h. Jets of gas that make up the solar chromosphere.
   i. The last glowing bead of light from the sun at the beginning (or the first one at the end) of a total solar eclipse.
   j. A small (earth-sized), dense (tons per cubic centimeter), under-luminous star, located to the lower left of the main sequence on a color-magnitude diagram.

Part II

1. a.   b.

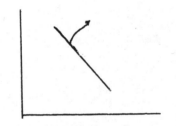

   c. IRAS detected infrared radiation from dark clouds called globules, indicating that stars may be formed inside them. An IRAS scan of the Andromeda galaxy revealed similar radiation in a ring-shape and another infrared source near the center. Observations of the Tarantula Nebula in the Large Magellanic Cloud showed many recently formed stars, with different wavelengths of radiation representing different stages of evolution.

2. a.

γ-rays	x-rays	uv	visible	ir		radio	
1A	100A	1μ			1 mm		1 m

   b.

   ⌐6000 K
   ⌐3500 K

   c. Wein's displacement law shows in the shift of the peak to
shorter wavelengths for higher temperatures; the
Stefan-Boltzmann law shows in the great increase (proportional
to the fourth power of temperature) of the curve for 6000 K.

3. a. M stars. Lines of molecules, especially titanium oxide.
   b. G stars. Hydrogen lines are visible; strong H and K lines are
present.
   c. O stars. Hydrogen spectral lines are weak. Ionized helium
lines are present.
   d. Normally we see absorption lines because as we look into the
solar atmosphere we see through cooler gas to a background of
hotter gas. Our view extends down to the photospheric layers
where the continuum is formed. When we see the chromosphere at
an eclipse, we see the gas superimposed on the sky. There is no
hotter background gas, and therefore we see the spectral lines
in emission.

4. a. The exit cone has its apex on the surface of a star that has
collapsed past the photon spghere but has not yet reached the
event horizon. It includes all the paths of light that escape
from the star.
   b. Radio.
   c. A given pulse reaches us at different times at different
frequencies. The more electrons that the signal has passed
through, the more the pulse is dispersed. Thus the dispersion
method tells us the distance to the pulsar.

5. a. Diagrams with deferent and epicycle for Ptolemaic method and
with one planet passing by another as they both circle the sun
at the center for Copernican method, as in Figure 2-5.
   b. First law: Planets circle the sun in ellipses, with the sun at
one focus.
Second law: Line joining planet and the sun sweeps out equal
areas in equal times.
Third law: Period squared = distance cubed.
   c. $P^2 = (0.4)^3 = 0.16 \times 0.4 = 0.064$ $P = .25$ year.

6. a. Photography, seismology, experimentation on the soil, infrared
measurements from orbit.
   b. In very brief summary, although the very first results from the
biology experiments seemed to indicate signs of life, further
results were more in line with the existence of certain unusual
chemical processes on Mars's surface. The finding of a chemical
experiment that no organic chemicals are present in Mars's soil
was a strong point against the interpretation of the original
results as life processes. Since the Viking results, work in
terrestrial laboratories has found chemical reactions that may
be the ones that were happening on Mars.
   c. $\Delta\lambda/\lambda = v/c = 40{,}000$ km/hr $\times$ 1/3600 sec/hr $\div$ 3×10^5 km/sec = 12
km/sec $\div$ 3×10^5 km/sec = $1/3 \times 10^4$. Thus $\Delta\lambda = 10$ cm $\div$ 3×10^4 =
3 cm/10^4 = 300 microns. Thus the new wavelength is 10 cm + 300

microns or 10.0003 cm.

7. a. Mercury, Venus, Earth, Mars, Jupiter, Saturn, Uranus, Neptune, Pluto.
    b. Jupiter, Saturn, Uranus, Neptune, Earth, Venus, Mars, Mercury, Pluto. (Order of giant planets shouldn't matter.)
    c. Jupiter, Saturn, Uranus, Neptune, Earth, Venus, Mars, Mercury, Pluto. (Order of giant planets shouldn't matter.)
    d. Meteoritic impacts, especially in the early solar system.
    e. Uranus essay should mention several points, perhaps including color, lack of contrast on surface, discovery of a few clouds and thus the rotation period, equal temperatures at both poles, offset and off-spin-axis magnetic field, craters on moons, strange terrain on Miranda, etc.

8. 6.4 and 3.2 solar masses.

Final Exam #2

1. a. The universe looks about the same in all directions and at all distances; that is, it is homogeneous and isotropic.
    b. The universe is homogeneous, isotropic, and unchanging in time. The perfect cosmological principle adds the statement about being unchanging in time. Since the universe is expanding, new matter must be continually being created to keep the density unchanging in time; this is part of the steady state theory.
    c. If we see the differences in the composition of the universe at any distance from us, then the steady-state theory is ruled out.
    d. The black-body radiation. This radiation has been detected as coming uniformly from all directions, and being equivalent to radiation from a black body at a temperature of 3 K. It is the remnant of the primeval fireball, and has been redshifted from the high temperature that it once was when the universe became transparent, to only 3° above absolute zero today. Its isotropy and its black-body shape lead us to believe that it comes from the big bang itself.

2. a. Quasars were discovered by comparison of observations in the optical and radio parts of the spectrum. It was the pinpointing of radio observations and the search for optical objects at those locations that led to their identification. The break-through was the realization that their unusual spectra were highly redshifted.
    b. If the quasars follow Hubble's law their tremendous redshifts indicate that they must be at very great distances. For us to be able to see them at all, they must then be such prodigious radiators of energy that we cannot explain how they could do so. Thus, some scientists prefer to think that the quasars are relatively close by and physically associated with other objects that are known to be closer to us. The validity of this reasoning depends on statistical arguments.
    c. The energy problem might still exist because we have to account for the enormous redshifts. If they are caused by Doppler

shifts, great amounts of energy are necessary to provide the high velocities. These statements have been challenged, and it has also been held that the energy problem would be greatly reduced.

d. The quasars may be a step in a progression of galaxies with bright nuclei. The "fuzz" detected by searching for traces of arms around quasi-stellar objects adds credence to the link between quasars and galaxies.

3. a. Very Large Array. The interferometric array in New Mexico, containing 27 antennas stretched across "Y" 27-km across.
   b. Very Long Baseline Interferometry.
   c. Quicker measurements, quicker reduction of data, and greater sensitivity.
   d. Higher resolution.
   e. Radio.
   F. Head-tail galaxies, small size of water vapor radio sources, and giant radio galaxies.

4. a. Closed, with a final collapse into a "big crunch." Open, with an ever-expanding universe.
   b. The amount of deuterium is very sensitive to the density of the universe at the time when the deuterium was formed. The present abundance of deuterium can thus tell us whether or not there was sufficient density, and thus matter present, to "close" the universe eventually.
   c. At present, there seems to be enough deuterium for the universe to have a relatively low density. This evidence suggests there may not be enough matter to close the universe.
   d. If we look far enough in any direction, we should eventually see the bright surface of a star, so the night sky should have the uniform brightness of the surface of a star. But the night sky is dark. The solutions to this paradox include the expansion of the universe, which causes photons of light travelling towards us to lose energy through the redshift, and the notion that most of the stars (or galaxies) that would be sending us light for the sky to be bright would be so far away that we would be looking back a long way in time, to a period before those necessary stars even existed.

5. a. $10 \text{ mm} \times 20 \text{ A/mm} = 200 \text{ A}$.
   b. $v = c\Delta\lambda/\lambda = 3 \times 10^5 \text{ km/sec} \times 2 \times 10^2 \text{ A} \div 6 \times 10^3 \text{ A} = 6 \times 10^7 \div 6 \times 10^3 = 10^4 \text{ km/sec}$.
   c. $\Delta\lambda/\lambda = 200/6000 = 1/30$. For $\lambda = 4000 \text{ A}$, $\Delta\lambda = 4000/30 = 130 \text{ A}$. New wavelength $= 4000 + 130 = 4130 \text{ A}$.
   d. $v = Hr$, $r = v/H = 10^4 \text{ km/sec} \div 50 \text{ km/sec/Mpc} = 10^4/50 \text{ Mpc} = 2 \times 10^2 \text{ Mpc}$.

6. a. A planetary nebula.
   b. An absorption and emission nebula.
   c. Reflection nebulae.
   d. Because the abundances of these molecules had not been expected to be as high as they were; thus it may be easier to form molecules in space than had been thought. Also, masering had

not previously been suspect.

e. On dust grains, especially in high-density clouds protected by dust from strong radiation (e.g. Orion Molecular Cloud).

7. a. Spiral. Because we are inside it, looking along the plane of the galaxy.

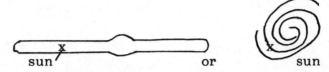

b. Quasars were discovered when they were pinpointed as the optical objects that correspond to some strong radio sources. They are strange because they have such high redshifts that by Hubble's law they are the farthest objects we can detect, and because to be as bright as they are even at that distance, they must be giving off incredible amounts of energy. They can be identified as quasars by their spectra or, for a preliminary screening, by their ultraviolet excesses.

c. A pair of quasar images separated in the sky by only 6 arc sec. High-resolution optical images have now shown an intervening galaxy that is acting as a gravitational lens. This is the first detection of a gravitational lens.

ANSWERS TO SAMPLE EXAMINATIONS

Final Exam #3
Multiple-Choice and Fill-in

1. a. different from
   b. mascons
   c. like
   d. more massive

2. b

3. a. higher
   b. molten
   c. both

4. a. false
   b. false
   c. false

5. a. sometimes in the daytime
   b. near the sun in the sky

6. a. seismometers
   b. corner reflectors

7. a. slowly
   b. square
   c. cube
   d. 85 years
   e. none

8. a. volcanoes
   b. polar caps
   c. water
   d. has not

9. Jupiter, Saturn, Uranus

10. a. parallaxes
    b. Cepheid variables
    c. RR Lyrae variables
    d. the brightest object in each
       cluster of galaxies is of
       the same magnitude.

11. a. Jupiter
    b. Saturn
    c. asteroid belt
    d. Uranus

12. a. gas, dust, and rock
    b. false
    c. 100 million km
    d. Halley's
    e. 1986

    f. 2062

    g. false

13. a. wide
    b. narrow
    c. wide

14. a. gamma rays, x-rays, visible, infrared, radio waves
    b. x-rays, gamma rays (ultraviolet okay too)

15. a.

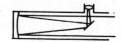

    or

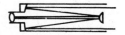

    b. A paraboloid focuses parallel light to a point.
    c. refracting
    d. The mirror doesn't have to be perfect, the mirror can be supported all along its back so it doesn't sag.

16. a. Pleiades
    b. Crab Nebula
    c. dying

17. 5-meter

18. a. 11 years
    b. do

19. a. radio
    b. number in a catalogue
    c. quasar

20. a. G
    b. 6000
    c. red giant
    d. 3000 K
    e. white dwarf
    f. 10,000
    g. red giant

21. emission line

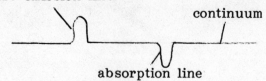

continuum

absorption line

22. almost like a star

23. a. black hole
    b. mass
    c. 5
    d. x-ray

24. a. infrared
    b. infrared
    c. x-ray
    d. ultraviolet
    e. visible

25. a. quasars
    b. faster than
    c. does not
    d. Einstein

26. a. velocity
    b. distance
    c. false
    d. Schmidt
    e. does not prove

27. Barnard's star, pulsars,
       center of our galaxy,
       Andromeda galaxy, quasars.

28. a. false
    b. true

29. a. newly brightened
    b. white dwarf
    c. a giant companion

30. a. H II regions
    b. open clusters
    c. O and B stars
    d. being born

31. a. hydrogen
    b. atom
    c. further
    d. false
    e. deuterium
    f. lower

32. a. open
   b. young
   c. reflection
   d. large

33. a. CO
   b. H_2O
   c. ammonia
      (See list in Appendix 11.)
   d. longer
   e. more accurate
   f. temperatures
   g. densities

34. a. emission
   b. H II region
   c. emission
   d. absorption nebula

35.

     H   D

36. The nuclear bulge, the disk,
   the galactic halo, the galactic
   corona.

37. Sa, Sb, Sc. Elliptical has
   no arms.

38. a. 3° background radiation
   b. the big bang
   c. big-bang

39. a. cannot
   b. does not
   c. emission lines

40. galaxies, black holes,
   neutron stars, white dwarfs,
   planets

41. black holes, neutron stars,
   white dwarfs, planets,
   galaxies

42. a. hotter
   b. smaller

43. a. radio
   b. neutron
   c. Crab

44. a. radio
    b. optical
    c. x-ray
    d. 1/1,000 sec
    e. supernovae

45. a. water vapor
    b. cooler
    c. the galactic center

46. a. shorter
    b. High-Energy Astronomy Observatory
    c. Advanced X-ray Astrophysics Facility

47. a. neutron star
    b. disk of matter
    c. precession of the object

48. a. radio
    b. resolution

49. a. event horizon
    b. mass
    c. high

50. a. chromosphere
    b. corona
    c. new
    d. year